Serial Textbooks of Ethiopian ATVET Supported by the Aid Program of Chinese Government

BEE KEEPING

China Agriculture Press

EDITORIAL BOARD

1. Leader Committee

Leaders in the Ministry of Agriculture and Rural Affairs (MARA), P. R. China

Qu Dongyu	Ministry of Agriculture and Rural Affairs, MARA, P. R. China	Deputy Minister
屈冬玉	中华人民共和国农业农村部	副部长
Sui Pengfei	Department of International Cooperation, MARA, P. R. China	Director-General
隋鹏飞	中华人民共和国农业农村部国际合作司	司长
Ma Hongtao	Department of International Cooperation, MARA, P. R. China	Deputy Director-General
马洪涛	中华人民共和国农业农村部国际合作司	副司长
Wu Changxue	Department of International Cooperation, MARA, P. R. China	Division Director
吴昌学	中华人民共和国农业农村部国际合作司	处长

Leaders in the Embassy of the People's Republic of China in Ethiopia

La Yifan	Embassy of the People's Republic of China in Ethiopia	Ambassador
腊翊凡	中华人民共和国驻埃塞俄比亚大使馆	大使
Liu Yu	Embassy of the People's Republic of China in Ethiopia	Counsellor
刘峪	中华人民共和国驻埃塞俄比亚大使馆	参赞
Luo Pengcheng	Embassy of the People's Republic of China in Ethiopia	Secretary
罗鹏程	中华人民共和国驻埃塞俄比亚大使馆	秘书

Zhu Wenqiang 朱文强	Embassy of the People's Republic of China in Ethiopia 中华人民共和国驻埃塞俄比亚大使馆	Secretary 秘书

Leaders in the Ministry of Agriculture and Natural Resources of Ethiopia

Tesfaye Mengiste	Ethiopian Ministry of Agriculture and Natural Resources	State Minister
Wondale Habtamu	Extension Directorate of Ethiopian Ministry of Agriculture and Natural Resources	Director-General
Kebede Atiseb	Agriculture Advisory and Training Directorate of Ethiopian Ministry of Agriculture and Natural Resources	Director
Kebede Beyecha	Federal Alage ATVET College	Dean
Chala Feyera	Federal Alage ATVET College	Vice Academic Dean
Tamirat Tesema	Federal Agarfa ATVET College	Dean
Akele Molla	Federal Agarfa ATVET College	Vice Academic Dean
Debela Bersisa	Holeta TVET College	Dean
Getachew Cibsa	Holeta TVET College	Outcome-based Training Vice Dean

2. Author Committee

Chief Author:

Tong Yu'e 童玉娥	Center of International Cooperation Service, MARA, P. R. China 中华人民共和国农业农村部国际交流服务中心	Director-General 主任

Associate Authors:

Chinese Associate Authors:

Luo Ming	Center of International Cooperation Service, MARA, P. R. China	Deputy Director-General
罗鸣	中华人民共和国农业农村部国际交流服务中心	副主任
Lin Huifang	Center of International Cooperation Service, MARA, P. R. China	Deputy Director-General
蔺惠芳	中华人民共和国农业农村部国际交流服务中心	副主任
Wang Jing	Center of International Cooperation Service, MARA, P. R. China	Division Director
王静	中华人民共和国农业农村部国际交流服务中心	处长
Guo Su	Center of International Cooperation Service, MARA, P. R. China	Deputy Division Director
郭粟	中华人民共和国农业农村部国际交流服务中心	副处长
Li Ronggang	Jiangsu Internet Agriculture Instructing Center, P. R. China	Professor
李荣刚	江苏省互联网农业指导中心	推广研究员
	Term 15, 16, 17 Chinese ATVET Project to Ethiopia	Coordinator
	援埃塞俄比亚农业职教组第15、16、17期	协调员

Ethiopian Associate Authors:

Ayele Gizachew	Agriculture TVET Office of Ethiopian Ministry of Agriculture and Natural Resources	Former Office Head
Getachew Demisie	Agriculture TVET Office of Ethiopian Ministry of Agriculture and Natural Resources	Office Head

Coauthors:
Chinese Coauthors:

Ma Lina	Rural Development Bureau of Xindu District Chengdu City, Sichuan Province, P. R. China	Senior Livestock Engineer
马俪娜	四川省成都市新都区农村发展局	高级畜牧师
	Term 14, 15, 16, 17 Chinese ATVET Project to Ethiopia	Livestock Expert
	援埃塞农业职教组第 14、15、16、17 期	畜牧专家
Duan Zhenhua	Center for Animal Disease Prevention and Control of Yulin City, Guangxi Zhuang Autonomous Region, P. R. China	Veterinarian
段振华	广西壮族自治区玉林市动物疫病预防控制中心	兽医师
	Term 16 Chinese ATVET Project to Ethiopia	Veterinary Expert
	援埃塞农业职教组第 16 期	兽医专家
Hu Zuobin	Wusheng County Bureau for Livestock and Food of Sichuan Province, P. R. China	Senior Veterinarian
胡佐斌	四川省武胜县畜牧食品局	高级兽医师
	Term 13, 14, 15, 16 Chinese ATVET Project to Ethiopia	Veterinary Expert
	援埃塞农业职教组第 13、14、15、16 期	兽医专家
Hu Xiaoquan	Honghuagang District Agricultural Bureau of Zunyi City, Guizhou Province, P. R. China	Senior Veterinarian
胡小全	贵州省遵义市红花岗区农业局	高级兽医师
	Term 13, 15, 17 Chinese ATVET Project to Ethiopia	Veterinary Expert
	援埃塞农业职教组第 13、15、17 期	兽医专家

Ethiopian Coauthors:

Girma Tirfessa	Animal Science Department, Alage ATVET College, Zeway, Ethiopia	Animal Science (BSC)
Kinfu Guta	Animal Science Department, Alage ATVET College, Zeway, Ethiopia	Animal Production and Technology (BSC)
Habtom Negussie	Animal Science Department, Alage ATVET College, Zeway, Ethiopia	Animal Science (MSC)
Bekele Abdissa	Animal Science Department, Agarfa ATVET College, Robe, Ethiopia	Animal Production(BSC)
Cheru Tesfaye	Animal Science Department, Agarfa ATVET College, Robe, Ethiopia	Animal Production(BSC)
Milion Bulo	Agriculture TVET office of Ethiopian Ministry of Agriculture and Natural Resources	Expert

3. Reviser Committee

Chinese Revisers:

Yang Yang	Center of International Cooperation Service, MARA, P. R. China	Programme Officer
杨飏	中华人民共和国农业农村部国际交流服务中心	项目官员
Fu Yan	Center of International Cooperation Service, MARA, P. R. China	Programme Officer
付严	中华人民共和国农业农村部国际交流服务中心	项目官员
Zhou Min	Center of International Cooperation Service, MARA, P. R. China	Programme Officer
周敏	中华人民共和国农业农村部国际交流服务中心	项目官员
Li Jun	Center of International Cooperation Service, MARA, P. R. China	Programme Officer
李俊	中华人民共和国农业农村部国际交流服务中心	项目官员

Ethiopian Reviser:

Abraham Zewdie	Ethiopian Ministry of Livestock and Fishery	Expert

FOREWORD

Ethiopia is the second-most populous nation on the African continent and houses the headquarters of the African Union. With the agriculture and livestock sector playing a pivotal role in its national economy, Ethiopia boasts many advantages in developing agriculture. China and Ethiopia have established diplomatic relations for almost half a century. Building on the profound traditional friendship, the two countries have generated fruitful outcomes in cooperation in various fields. In particular, China-Ethiopia agricultural cooperation is featured by many highlights as follows:

The China-Ethiopia Agricultural Technical Vocational Education and Training Program (the ATVET Program), as a key program in China-Ethiopia cooperation, was jointly launched by the Ministries of Agriculture of China and Ethiopia in 2001 and transformed into a foreign aid program of the Chinese government in 2012. Up to now, the Chinese government has dispatched 405 instructors in 16 batches to Ethiopia for the program. These Chinese instructors have devoted themselves to Ethiopia for 15 years. Stationed in 13 ATVET colleges, they have given lectures on 56 subjects in 5 disciplines, namely plant science, animal science, natural resources, veterinary science and agricultural cooperatives. Thanks to their dedication, 2,100 local teachers, 13,000 agricultural technicians and 39,000 students have been trained and imparted with over 70 practical and advanced technologies. These Chinese teachers also helped Ethiopia establish a tailor-made agricultural vocational education system. This program has won recognition from the Ministry of Agriculture and Natural Resources of Ethiopia and the beneficiaries, the warm welcome by local students, teachers and farmers, and extensive coverage by Chinese and foreign core media. It is undoubtedly a role model of China's pro-Africa agricultural cooperation.

After years' of joint study and exploration, Chinese instructors and Ethiopian partners have realized that the absence of professional textbooks is a serious constraint to the sustainable development of Ethiopian agricultural vocational education system. Therefore, the Chinese and Ethiopian governments attach great importance to the development of textbooks. The two countries conducted a range of studies and drafted a textbook compilation and publishing plan. According to the plan, the textbook series consist of 13 volumes, involving poultry production, animal health, horticulture, small-scale irrigation, dairy production, apiculture, water and soil conservation, beef cattle, field crop, coffee, tea and spice crops, cotton, forestry, farm machinery, etc. Building upon the experience of the ATVET program and taking into account the practical demand of agricultural development in Ethiopia, the textbooks have summarized the advanced agricultural theories and technologies of China and will serve as practical guidance to more agricultural practitioners in Ethiopia. Authors of the textbooks are long-term practitioners in the ATVET program and have gone through rigorous review by the Leader Committee. In order to make sure that the textbooks are localized and operable, authors have had thorough communication with the Ethiopian Ministry of Agriculture and Natural Resources, received overall guidance from the Leader Committee and worked wholeheartedly to develop and revise each and every volume.

The publication of the textbooks have obtained strong support from the Chinese Ministry of Agriculture and Rural Affairs, Chinese Ministry of Commerce, Embassy of the People's Republic of China in Ethiopia, Ethiopian Ministry of Agriculture and Natural Resources and relevant Ethiopian vocational education colleges. The textbooks will certainly provide effective means for the training of practical agricultural professionals in Ethiopia, promote the sustainable development of Ethiopian agricultural vocational education, and enhance the capacity and level of agricultural development in Ethiopia.

PREFACE

Apiculture or beekeeping is the art of managing bees with the intention of getting the maximum return from this work with the minimum of expenditure. It is one of the agricultural subsectors that most suits the rural poor and also contributes significantly to income diversification for those who are better off. It is simple and relatively cheap to start, as it requires low level of inputs (land, labor and capital). The bee products and by-products supply income that contributes to the improvement of the livelihood of the rural people. Besides, the role played by honeybees in food security by increasing the productivity of crop production through intensifying pollination is already an established fact. Beekeeping is exceptionally sustainable as the activity has no impact on the environment and rather it stabilizes fragile areas and helps in reclaiming degraded lands and increases biodiversity.

Ethiopia has a longstanding beekeeping practices and is also endowed with huge apicultural resources and it has been an integral part of other agricultural activities, where about one million households have kept honeybees since 2007. Data from CSA (2013) indicated that more than 5.15 million hived honeybee colony populations were found in the country. From this total hives, the greater part (96.38 percent) was reported to be traditional. As a result, the estimate of total honey production was about 53,680 – ton. This made Ethiopia a leading in Africa and tenth in the world in honey production, respectively. Similarly, it stood first in Africa and third in the world, in beeswax production (FAOSTAT, 2005).

The majority of the honey and beeswax produced serves as cash crop and is used as a tool to increase income and livelihood improvement. Beekeeping also contributes to the country's economy through foreign exchange earnings. Yet, about 97% of the production system was traditional beekeeping system using basket hives made from different local materials in various shapes and volumes (CSA, 2013). This was one of the important factors that caused the country not to exploit the enormous potential it had in beekeeping. It was therefore, justifiable to support this important and potential subsector through institutionally organized research development and training endeavors at all levels of production, processing and marketing (Gemechis et al., 2012). With this regard, beekeeping training have been carrying out for several years to earn excellent hive products and adoption of modern beeping technologies in ATVETs, governmental institution and farmer training centers without uniformity, so that they have different guidelines, standards and specification to accomplish their training methods. This book aims to maintain uniformity of the training operation in agricultural TVET college throughout the country.

The book consists of 11 chapters in which all are presented in detail in-line with the Ethiopian Occupational standards (EOS). Module 1 covers the identification and characterization of honey bee species and races and their behaviors. Module 2 deals about honeybee colonies and their organization. In module 3, anatomy and physiology of honeybees is illustrated. Module 4 enables to identify and prepare beekeeping equipment and tools. Module 5 helps to establish an apiary and identify honeybee flora and bee toxic plants and develop floral calendar. Module 6 covers the routine activities to be undertaken in beekeeping. Module 7 enables to perform honeybee colony management practices. Module 8 enables to harvest, process and store honeybee products. Module 9 deals about queen bee rearing. Module 10 enables to identify honeybee diseases, pest and predators, and their treatment and control measures. Finally, module 11 enables to identify and develop beekeeping records.

CONTENTS

FOREWORD
PREFACE

MODULE 1: IDENTIFY HONEYBEE SPECIES, RACES AND THEIR BEHAVIOR 1

INTRODUCTION 1
1 HONEY BEE SPECIES AND THEIR BEHAVIOR 1
  1.1 Common Species of Honeybees 2
    1.1.1 *Apis mellifera* 2
    1.1.2 *Apis cerana* (Oriental Honeybee) 3
    1.1.3 *Apis florea* (Dwarf Honeybee) 4
    1.1.4 *Apis dorsata* (Giant Honeybee) 5
  1.2 Other New Honeybees 6
    1.2.1 *Apis andreniformis* (Black Dwarf Honeybee) 6
    1.2.2 *Apis koschevnikovi* 6
    1.2.3 *Apis laboriosa* 7
    1.2.4 *Apis nigrocincta* 7
    1.2.5 *Apis nuluensis* 7
  1.3 Stingless Bees 8
2 RACES OF HONEYBEES AND THEIR BEHAVIOR 9
  2.1 Distinctive Characteristics of Races of Bees 10
  2.2 European Honeybee Races 11
    2.2.1 *Apis mellifera ligustica* (Italian Bees) 11
    2.2.2 *Apis mellifera mellifera* (Dark Bees) 11
    2.2.3 *Apis mellifera carnica* (Carniolan Bees) 12
    2.2.4 *Apis mellifera caucasica* (Russian Bees) 12
  2.3 African Honeybee Races 13
    2.3.1 *Apis mellifera intermissa* 13
    2.3.2 *Apis mellifera saharansis* (Saharan Bees) 14

    2.3.3 *Apis mellifera lamarckii* (Egyptian Bees) 14
    2.3.4 *Apis mellifera jemenitica* 14
    2.3.5 *Apis mellifera scutellata* 14
    2.3.6 *Apis mellifera litorea* 14
    2.3.7 *Apis mellifera monitcola* 15
    2.3.8 *Apis mellifera adansonii* 15
    2.3.9 *Apis mellifera unicola* 15
    2.3.10 *Apis mellifera capensis* (Cape Town Bees) 15
  2.4 Differences between African and European Honey Bees 15
  2.5 Ethiopian Honeybee Races 17
    2.5.1 *Apis mellifera monticola* 17
    2.5.2 *Apis mellifera bandansii* 17
    2.5.3 *Apis mellifera scutelltata* 17
    2.5.4 *Apis mellifera jemenitica* 18
    2.5.5 *Apis mellifera woyi-gambella* 18
  SELF-CHECK QUESTIONS 18
  REFERENCES 20

MODULE 2: RECOGNIZE HONEY BEE COLONY AND ITS ORGANIZATION 21

INTRODUCTION 21

1 HONEY BEE CASTES AND THEIR DUTIES IN THE COLONY 21
  1.1 Queen Bee 22
  1.2 Worker Bees 23
  1.3 Drone Bees 26

2 DEVELOPMENT STAGES OF BEES 26
  2.1 Egg 26
  2.2 Larva 27
  2.3 Pupa 29

3 COMMUNICATION SYSTEM OF HONEY BEES 29
  3.1 Pheromones 29
  3.2 Dances 30

  SELF-CHECK QUESTIONS 32
  REFERENCES 33

MODULE 3: IDENTIFY THE STRUCTURE, LOCATION AND FUNCTION OF HONEYBEE'S BODY PARTS 34

INTRODUCTION 34

1 COMPONENTS AND FUNCTIONS OF EXTERNAL BODY PARTS 34

1.1	Head Region	34
	1.1.1 Antennae	36
	1.1.2 Eyes	36
	1.1.3 Mouth	37
1.2	Thorax Region	38
	1.2.1 Wings	38
	1.2.2 Legs	39
1.3	Abdomen Region	41
	1.3.1 Wax Scales	41
	1.3.2 Spiracles	41
	1.3.3 Sting	41
1.4	Exoskeleton of Honey Bees	42
1.5	Body Hair	43
2	**COMPONENTS AND FUNCTIONS OF INTERNAL BODY PARTS**	**43**
2.1	Digestive Organs and System	43
2.2	Respiratory Organs and System	44
2.3	Circulatory System	44
2.4	Reproductive System	45
2.5	Nervous System	45
2.6	Endocrine/Gland System	46
	SELF-CHECK QUESTIONS	**47**
	REFERENCES	**50**

MODULE 4: IDENTIFY AND PREPARE BEEKEEPING EQUIPMENTS AND TOOLS 51

INTRODUCTION 51

1 BEE HIVES AND THEIR ACCESSORIES 51

1.1	Traditional Hive (Fixed Comb Hives)	51
1.2	Top-bar (Transitional) Hives	54
1.3	Modern Hives (Improved Frame Hives)	58
2	**EQUIPMENT FOR ROUTINE ACTIVITIES AND COLONY MANIPULATION**	**61**
2.1	Materials for Foundation Sheet Making	61
2.2	Feeds and Feeder	61
2.3	Queen Rearing Equipment	62
2.4	Farm Tools	62
2.5	Honey Harvesting and Processing Equipment	63
	2.5.1 Honey Harvesting Equipment	63
	2.5.2 Honey Processing Equipment	64

2.6 Personal Protective Cloth	65
SELF-CHECK QUESTIONS	66
REFERENCES	67

MODULE 5: ESTABLISH APIARY 68

INTRODUCTION 68
1 OBTAIN SWARMS AND HIVES 68
2 SELECTING AN APIARY SITE 69
3 CLEARING AND FENCING THE SITE 70
   3.1 Clearing the Site 70
   3.2 Fencing the Site 70
4 BAITING AND COLONIZATION OF THE HIVE 71
   4.1 Materials for Baiting Swarm 71
   4.2 Procedure to Baiting Swarm 72
5 INSTALL BEE HIVE AND HIVE STAND 72
   5.1 Hive Hanging 73
   5.2 Install Hive Stand 74
6 IDENTIFY HONEYBEE FLORA 76
   6.1 Honey Plants and Pollen Plants 77
   6.2 Floral Calendars 78
   6.3 Assessment of Areas for Honeybee Flora 79
   6.4 Seasonal Blooming Honey Plants for Bee Forage 80
      6.4.1 Common Honeybee Flora in Ethiopia 81
      6.4.2 Toxic Plants for Bees 85
SELF-CHECK QUESTIONS 86
REFERENCES 88

MODULE 6: CARRY OUT ROUTINE BEE KEEPING ACTIVITY 89

INTRODUCTION 89
1 HANDLING BEES 90
2 MAINTAINING CLEAN HIVES AND APIARY 90
3 ADMINISTRATION ACTIVITIES 91
4 MOVING HIVED COLONY 92
   4.1 Long Distance Moving Hived Colony 92
   4.2 Short Distance Moving Hived Colony 93
5 WEIGHING HIVED COLONY 93
6 MAKE COMB FOUNDATION SHEET AND FIX TO FRAME 94
   6.1 Materials Required for Making Foundation Sheet 95

    6.2 Procedure in Making Foundation Sheet ⋯⋯ 95
    6.3 Fixing Foundation Sheets into Frames ⋯⋯ 96
7 **PREPARE SMOKER FOR USE** ⋯⋯ 96
8 **OPEN AND REASSEMBLE THE HIVE** ⋯⋯ 97
9 **INSPECT THE COLONY** ⋯⋯ 97
    9.1 General Guidelines to Inspect ⋯⋯ 98
    9.2 External Inspection ⋯⋯ 99
    9.3 Internal Inspection ⋯⋯ 99
        9.3.1 Requirements for Internal Hive Inspection ⋯⋯ 99
        9.3.2 Internal Hive Inspection Procedure ⋯⋯ 100
        9.3.3 Reading Frames or Top-bars ⋯⋯ 100
SELF-CHECK QUESTIONS ⋯⋯ 101
REFERENCES ⋯⋯ 102

MODULE 7: MANAGING BEE COLONY ⋯⋯ 103

INTRODUCTION ⋯⋯ 103
1 **PROVIDING SUPPLEMENTARY FEED** ⋯⋯ 103
    1.1 Time of Feeding Honey Bees ⋯⋯ 104
    1.2 Categories of Feeding Techniques ⋯⋯ 105
    1.3 Feed Types for Feeding Honey Bees ⋯⋯ 105
        1.3.1 Honey ⋯⋯ 105
        1.3.2 Sugar Syrup ⋯⋯ 106
        1.3.3 Pollen ⋯⋯ 107
        1.3.4 Pollen Substitute ⋯⋯ 108
        1.3.5 Pollen Supplement ⋯⋯ 108
2 **DIVIDING AND UNITING COLONIES** ⋯⋯ 108
    2.1 Dividing a Colony ⋯⋯ 108
    2.2 Uniting Colonies ⋯⋯ 110
3 **MIGRATING COLONIES** ⋯⋯ 111
4 **SUPERING AND REDUCING THE HIVE** ⋯⋯ 112
    4.1 Supering ⋯⋯ 112
    4.2 Reducing Super ⋯⋯ 113
5 **CONTROLLING COLONY SWARMING** ⋯⋯ 114
    5.1 Contributing Factors for Swarming ⋯⋯ 115
    5.2 Preventing Swarming of Honey Bees ⋯⋯ 115
6 **PREVENTIONS OF ABSCONDING** ⋯⋯ 116
7 **INSERTING AND REMOVING QUEEN EXCLUDER** ⋯⋯ 117
8 **TRANSFERRING BEE COLONY** ⋯⋯ 118

8.1　Season (Time) of Transferring ·· 118
8.2　Follow up after Transferring ·· 119
SELF-CHECK QUESTIONS ·· 119
REFERENCES ·· 120

MODULE 8: HARVEST, PROCESS AND STORE BEE PRODUCTS ·· 121

INTRODUCTION ·· 121

1　IDENTIFY AND CHARACTERIZE BEE PRODUCTS ·· 121

1.1　Honey ·· 121
   1.1.1　Chemical Composition and Physical Properties of Honey ·· 122
   1.1.2　Uses of Honey ·· 125
   1.1.3　Honey Quality and Categories ·· 125
1.2　Bees Wax ·· 126
   1.2.1　Wax Properties and Composition ·· 126
   1.2.2　Uses of Wax ·· 127
1.3　Propolis ·· 127
   1.3.1　Properties and Composition ·· 128
   1.3.2　Uses of Propolis ·· 128
1.4　Pollen ·· 128
   1.4.1　Properties and Composition ·· 129
   1.4.2　Uses of Pollen ·· 129
1.5　Royal Jelly or Bee Milk ·· 130
   1.5.1　Properties and Composition ·· 130
   1.5.2　Uses of Royal Jelly ·· 130
1.6　Bee Venom ·· 130

2　HARVEST AND PROCESS HONEY ·· 131

2.1　Honey Harvesting Seasons, Indicators and Requirement ·· 131
   2.1.1　Honey Harvesting Seasons ·· 131
   2.1.2　Indications for Honey Harvesting ·· 132
   2.1.3　Honey Harvesting Requirements ·· 132
2.2　Honey Harvesting Procedures ·· 133
2.3　Honey Extraction, Processing and Packaging ·· 134
   2.3.1　Honey Extraction ·· 135
   2.3.2　Honey Processing ·· 138
   2.3.3　Honey Packaging ·· 138
   2.3.4　Honey Storing ·· 139

3　PROCESS OLD AND BROKEN COMBS IN TO CLEAN WAX ·· 140

3.1　Straining Method of Wax Processing ·· 140

    3.2　Solar Wax Extraction Method ································· 141
    3.3　Steam Wax Extraction ··· 141
 4　COLLECT AND STORE POLLEN GRAIN, PROPOLIS, BEE VENOM AND
    ROYAL JELLY ··· 142
    4.1　Pollen Collection and Storage ································· 142
        4.1.1　Harvesting ·· 143
        4.1.2　Storage ··· 143
    4.2　Propolis Collection and Storage ······························ 143
    4.3　Royal Jelly Harvesting ·· 144
    4.4　Bee Venom Collection and Storage ··························· 144
        4.4.1　Production ·· 144
        4.4.2　Preparation ··· 145
    SELF-CHECK QUESTIONS ··· 145
    REFERENCES ·· 147

MODULE 9: REARING QUEEN BEES ····································· 148

INTRODUCTION ·· 148
1　THE PURPOSE OF QUEEN BEE REARING ····························· 149
2　BASIC REQUIREMENTS FOR QUEEN BEE REARING ·················· 149
    2.1　Breeding Stock ·· 149
    2.2　Sufficient Food (Nectar and Pollen) ·························· 151
    2.3　Sufficient Drone Bees ··· 151
    2.4　Suitable Weather ·· 151
3　CONDITIONS UNDER WHICH BEES RAISE THEIR OWN QUEENS ····· 151
    3.1　During Swarm Preparation ···································· 152
    3.2　When an Old Queen Fails to Lay Eggs ······················· 152
    3.3　When a Colony Lost Its Queen ······························· 152
4　METHODS OF QUEEN BEE REARING ··································· 152
    4.1　Grafting (Doolittle) Method ··································· 153
        4.1.1　Tools and Equipment Needed ····························· 153
        4.1.2　Technical Procedures of Grafting ························ 155
    4.2　Non Grafting Methods of Queen Rearing ····················· 160
        4.2.1　Miller Method ·· 160
        4.2.2　Hopkins Method ·· 161
        4.2.3　Alley Method ··· 163
        4.2.4　Splitting Method ·· 163
        4.2.5　Overcrowding Method ····································· 164
    SELF-CHECK QUESTIONS ··· 165

REFERENCES ... 166

MODULE 10: CARE FOR THE HEALTH OF HONEYBEE COLONY ... 167

INTRODUCTION ... 167

1 IDENTIFY AND CHARACTERIZE COMMON HONEYBEE DISEASES AND PARASITES ... 167
- 1.1 Bacterial Diseases ... 168
  - 1.1.1 American Foulbrood Disease ... 168
  - 1.1.2 European Foulbrood Disease ... 169
- 1.2 Viral Diseases ... 170
  - 1.2.1 Sac-brood Disease ... 170
  - 1.2.2 Virus Paralysis Disease ... 171
  - 1.2.3 Deformed Wing Virus ... 171
  - 1.2.4 Black Queen Cell Virus ... 172
- 1.3 Protozoal Diseases ... 172
  - 1.3.1 Amoeba ... 172
  - 1.3.2 Nosema Disease/Nosemosis ... 172
- 1.4 Fungal Diseases ... 174
  - 1.4.1 Chalkbrood Disease/Ascosphaerosis ... 174
  - 1.4.2 Stone-brood Disease ... 174
- 1.5 Parasites ... 175
  - 1.5.1 Varroa Mite/Varroatosis ... 175
  - 1.5.2 Tracheal Mite/Acaropisosis ... 176
  - 1.5.3 Bee Lice ... 176

2 PESTS AND PREDATORS ... 177
- 2.1 Pests of Honey Bee ... 177
  - 2.1.1 Wax Moth ... 177
  - 2.1.2 Ants and Termites ... 179
  - 2.1.3 Small Hive Beetle ... 180
  - 2.1.4 Other Insects ... 182
- 2.2 Predators of Honey Bees and Their Products ... 182
  - 2.2.1 Predatory Birds ... 182
  - 2.2.2 Man ... 182
  - 2.2.3 Mice ... 182
  - 2.2.4 Honey Badger ... 182
  - 2.2.5 Reptiles (Snake, Toads and Lizards) ... 183

3 PESTICIDE POISONING ... 184

4 PREVENT AND CONTROL HONEYBEE DISEASES AND PARASITES ... 185

	4.1	General Measures	185
	4.2	Traditional Methods	186
	4.3	Recommendations	187

5 DIAGNOSE AND TREAT INFECTED HONEYBEE COLONY 189

- 5.1 Characteristics of Healthy Honeybee Colony 189
- 5.2 Diagnosis and Treatment 191
  - 5.2.1 American Foulbrood Disease 191
  - 5.2.2 European Foulbrood Disease 192
  - 5.2.3 Viral Diseases 193
  - 5.2.4 Chalk-brood Disease 194
  - 5.2.5 Nosema 194
  - 5.2.6 Amoeba 196
  - 5.2.7 Varroa Mites 196
  - 5.2.8 Tracheal Mites 202
- **SELF-CHECK QUESTIONS** 203
- **REFERENCES** 205

MODULE 11: RECORD KEEPING 207

INTRODUCTION 207

1 RECORD – KEEPING PRACTICES IN BEEKEEPING 207

2 TYPES OF RECORD IN BEEKEEPING 208

- 2.1 Breeding Record 208
- 2.2 Colony Health Record 209
- 2.3 Colony Record 209
- 2.4 Production Record 209
- 2.5 Feed Record 210
- 2.6 Colony Source Record 210
- 2.7 Financial Record 211
- 2.8 Materials Record 211
- 2.9 Human Resource Record 211
- **SELF-CHECK QUESTIONS** 212
- **GLOSSARY** 213
- **ANSWER KEY OF SELF-CHECK QUESTIONS** 218

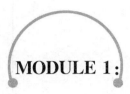

MODULE 1:
IDENTIFY HONEYBEE SPECIES, RACES AND THEIR BEHAVIOR

>>> INTRODUCTION

Honey bees are flying insects closely related to wasps and ants, which feed on pollen and nectar. They are a monophyletic lineage within the super family Apoidea, presently considered as a clade Anthophila (Danfort, 2006). Honeybees are available on every continent except Antarctica, in every habitat on the planet that contains insect-pollinated flowering plants. One genus of honeybees, which is called *Apis*, contains all the known species of honeybee subspecies. Bees in this genus can produce and store honey and build comb from wax. It is these properties that humankind has learned to use by learning how to manage colonies in hives. Different species of honeybees nest in one of two different ways, and this nesting behavior determines whether the bees can be kept inside a manufactured hive. Due to this reason, beekeepers should select the best species of bees that can suit in manmade hive.

There are several popular honeybee races that are raised in the world for honey and other purposes. Some of these races are considered gentler than others making them more ideal for beginning beekeepers. Each of them has their own advantage and disadvantage according to their original regional background.

Therefore, this module is developed to provide you the necessary information on identifying and characterizing common honeybee species and races, and their unique behaviors.

1 HONEY BEE SPECIES AND THEIR BEHAVIOR

Honeybee species in contrast with the stingless honeybee is any bee that is a member of the genus *Apis*, primarily distinguished by the production and storage of honey and the construction of perennial, colonial nests from wax. Honeybees are the only extant members of the tribe Apini, all in the genus *Apis*. Currently, around seven species of honeybee are recognized, with 44 subspecies. Some other types of related bees produce and store honey, but only members of the genus *Apis* are true honeybees (Michael, 1999).

> **Taxonomic classification of honey bees**
>
> Kingdom: Animalia
>   Phylum: Arthropoda
>   Class: Insecta
>   Order: Hymenoptera
>   Super family: Apoidea
>   Family: Apidae
>   Genus: *Apis*
>   Species: >20,000 spp.

Bees that produce enough honey to make harvesting worthwhile belong to two subfamilies honeybees (Apinae) and stingless bees (Meliponinae). Apinae has only one genus *Apis*, of which five species are economically important. These are *Apis florea*, *Apis dorsata*, *Apis cerana*, *Apis indica*, *Apis mellifera*, and *Apis laboriosa*.

Of these five species of honeybees, *A. mellifera* has greatest economic importance and widely distributed all over the world. As an inherited behavior characteristic, all honeybee colonies tend to store a certain amount of honey and pollen as their food reserve. The quantity of food stored depends upon several factors, including the seasonal availability of forage, the worker population of the colony and its rate of reproduction, the capacity of the nest, etc. Another important inherited behavior characteristic lies in the colony's natural site of comb construction. Some *Apis* species build single comb nests in the open; others build multiple-comb nests in dark cavities. The species of bees that build a series of parallel combs usually nest inside cavities, and this behavior enables them to nest inside manufactured containers.

1.1 Common Species of Honeybees

1.1.1 *Apis mellifera*

This honeybee is the most widespread species. It is native in Europe, the near and middle East of Asia and Africa. *A. mellifera* bees have been introduced in many other tropical regions. It is very dominant defensive behavior and tendency to quick swarm. One implication is that colonies should be kept away from houses and roads where people live or pass (Fig. 1).

This honeybee species constitutes an integral part of modern agricultural systems, furnishing crop pollination services as well as honey and beeswax. There are many geographical races of the common honeybee *Apis mellifera*, distributed widely throughout Europe, Africa, and parts of western Asia, as well as in the Americas.

There are about 25 races, of which the most important are: *A. mellifera ligustica*,

A. mellifera intermissa and *A. mellifera adasonii* (Segeren, 2004). All these races display similarities in their basic biological attributes, e. g. the construction of multiple-comb nests in dark cavities, colony social organization and division of labor, etc. In the wild, the natural nesting sites of *A. mellifera* are like those of *A. cerana*: caves, rock cavities and hollow trees. The nests are composed of multiple combs, parallel to each other, with a relatively uniform bee space. The nest usually has a single entrance.

The temperate races prefer nest cavities of about 45 liters in volume and avoid those smaller than 10, or larger than 100 liters. Colonies of the European races are composed of relatively large populations usually is range 15,000 – 60,000 (Segeren, 2004). *Apis mellifera* is the most productive of all honeybee species. Average honey yield of this honeybee species is 45 – 180kg per year in good honey yield area. It has high degree of adaptability. It is aggressive.

The serious disadvantage of this honeybee is its vulnerability to certain diseases especially a parasite like the varroa mite. Among the 25 races of *A. mellifera* in the world, most of them are in Africa.

1.1.2 Apis cerana (Oriental Honeybee)

The colony of *Apis cerana*, a typical honeybee, consists of several thousand female worker bees, one queen bee, and several hundred male drone bees. The colony constructed inside beeswax combs inside a tree cavity, with a special peanut-shaped structure on the margins of the combs where the queens are reared (Seeley, 2009). In *A. cerana*, reproductive swarming is similar to *A. mellifera*. *A. cerana* reproductive swarms settle 20 – 30m away from the natal nest (the mother or primary colony) and stay for a few days before departing for a new nest site after getting information from scout bees (Fig. 2).

As the wasps approach the entrance to the honeybee nest, more guard bees are alerted, which in turn increases their probability of being killed by heat-balling bees. Heat balling is a unique defiance system in which several hundred bees surround the wasp in a tight ball and vibrate their muscles in an effort to produce heat and effectively kill the wasp inside.

Alternatively, however, in the presence of a wasp, the bees may also just withdraw into their nests and await the heat-balling circumstances to develop naturally. Furthermore, other bees may just decide to fly away as an evasive measure in times of conflict, often altering their specific flight styles in order to avoid predation (Tan, 2007).

Apis cerana is quite similar to *A. mellifera* as it also nests in cavities, such as hives. Similar types of beekeeping can be done with both species. However, there are also important differences, especially in biology and behavior.

The most productive race is *A. cerana cerana* in China, least productive with much smaller colonies is *A. cerana indica* in India and Southeast Asia and *A. cerana javana*. It is useful to know that tropical *A. mellifera* and *A. cerana* races are smaller than European *A. mellifera* (Segeren, 2004).

This bee's range of distribution is far greater than that of *A. florea* and *A. dorsata*

. The normal nesting site is, in general, close to the ground, not more than 4 - 5 meters high. Average honey yield is 35kg/year (Pongthep, 1990).

Apis cerana has been kept in various kinds of hives, i. e. clay pots, logs, boxes, wall openings, etc. The several combs of an *A. cerana* colony are build parallel to each other, and a uniform distance. These bees can be kept in smaller hives. The honey production is not as that of *A. mellifera*. So now, it is replaced by strains of *A. mellifera* bees.

However, these two species of bees are difficult to keep them together in the same area because they cross breed each other without giving offspring. *Apis cerana* is tends to swarm and migrate to secure better food source. This honeybee species can resist to Nosema disease.

1.1.3 *Apis florea* (Dwarf Honeybee)

The name *florea* is a personal name of Romanian origin. *A. florea* is native to Southeast Asia, and therefore one of the most phylogenetically basal bees. This suggests that honeybees originated in this region, with *A. florea* being one of the oldest and potentially like ancestors; therefore, its first appearance in phylogeny occurs before that of *A. mellifera* and *A. cerana*. *A. mellifera* most likely evolved from *A. florea* and then migrated to Europe. *A. florea*, a dwarf honeybee, is a sister species of *Apis andreniformis* (Gupta, 2014).

In a single nest, there is high genetic diversity among the *Apis florea* bees. Since honey bee queens are polygamous, strong genetic variability exists. The tendency of this species to perform certain tasks is dependent on this variation. For example fanning of the nest is done by specific set of workers, when the nest reaches a specific temperature threshold (Jones, 2007).

A nest of *A. florae* consists of a single comb and nest in the open. Most nests are hung from slender branches of trees or shrubs covered with relatively dense foliage, usually from one to eight meters above the ground. To ward off ant attacks, the workers coat both ends of the nest support with sticky strips of propolis or plant gum, from 2.5 to 4cm wide. *A. florae* are the only honeybee that uses this defensive technique. During the season when there is an ample supply of nectar and honey, populous colonies of the dwarf honeybee send out multiple reproductive swarms. These colonies also have a high degree of mobility. Disturbance by natural enemies, exposure to inclement weather and scarcity of forage are among the major causes of colonies absconding.

In comparison with other honeybee species, the amount of honey that *Apis florae* workers will store in their nests is small, usually not exceeding several hundred grams per colony. Therefore, an *A. flora* is economically less important. Where nests of *A. florae* are abundant, several rural families can subsist on the income generated from bee hunting alone. Although the practice appears ecologically destructive, particularly insofar as it reduces a valuable population of natural pollinators, it does not always destroy the colony being hunted. Workers and laying queens of the dwarf honeybee are able to respond to nest predation quickly. The entire colony, accompanied by a laying queen, can fly several meters away to regroup, and later abscond. Some absconding colonies are able to survive to build their new combs in a nearby area. As its name implies, the dwarf honeybee is the smallest species of honeybee. Its size is about 7mm in length. Colony size of this honeybee species is small (about 5,000). It tolerates very hot temperature (up to 50℃) (Pongthep, 1990).

1.1.4 *Apis dorsata* (Giant Honeybee)

Apis dorsata found from the Indian subcontinent to Southeast Asia and the Philippines. The greatest populations of *Apis dorsata* are found include China, Indonesia, India, Pakistan, and Sri Lanka. They mostly reside in tall trees in dense forests, but also build nests on urban buildings. These bees are tropical and in most places, they migrate seasonally. Individual colonies migrate between nesting sites during the transition from the rainy to dry seasons and occupy each nesting site for about 3 - 4 months at a time (Paar, 2006). These bees build small combs that serve as temporary nests during their long migrations. *Apis dorsata* differs from the other bees in its genus in terms of nest design. Each colony consists of a single vertical comb made of workers' wax suspended from above, and the comb is typically covered by a dense mass of bees in several layers. The nests are mostly conical in shape and vary in size, reaching up to 1 meter in width.

Each cell within the comb is hexagonal in shape. *Apis dorsata* store their honey on the top right-hand corner of the comb and rear the worker and drone brood in the same location of the nest. Nests are constructed in the open and in elevated locations, such as on urban buildings or tall trees. These bees rarely build nests on old or weak buildings for safety concerns (Neupane, 2005). *Apis dorsata* can form dense aggregations at one nesting site, sometimes with up to 200 colonies in one tree. Each colony can have up to 100,000 bees and is separated by only a few centimetres from the other colonies in an aggregation (Paar, 2006). Some colonies also exhibit patterns of nest recognition, in which they return to the same nesting sites post migration.

Since the nests of *Apis dorsata* are fairly exposed and accessible to predators, they exhibit strong and aggressive defence strategies. Their predators include wasps, hornets, birds, and human bee-hunters. Their defense strategies typically include physical contact, especially when they face attacks from wasps. These giant honeybees utilize a method called 'heat balling', in which they heat their thoraces to a temperature of 45℃, which is lethal to wasps (Kastberger).

Another method that *Apis dorsata* utilizes against wasps is referred to as 'shimmering' behavior or defence waving. Bees in the outer layer thrust their abdomens 90° in an upward direction and shake them in a synchronous way. This may be accompanied by stroking of the wings. The signal is transmitted to nearby workers that also adopt the posture, thus creating a visible and audible 'ripple' effect across the face of the comb, in an almost identical manner to an audience wave at a crowded stadium. These wave-like patterns repel wasps that get too close to the nests of these bees and serve to confuse the wasp. In turn, the wasp cannot fixate on capturing one bee or getting food from the bees' nest, so the wasp will seek to find easier prey and leave this nest alone. Shimmering appears to be an evolutionary successful behavior for group living amongst social bees.

Workers of *A. dorsata* are able to fly at night, when the light of the moon is adequate. These bees are large in size 17 - 19mm, have 20,000 or more workers in colony, and produce honey up to 35kg/year (Pongthep, 1990). The colonies migrate up and down mountains to take advantage of seasonal food source, i.e. has nomadic nature. The workers *A. dorsata* are aggressive.

1.2 Other New Honeybees

In addition to above mentioned honeybees, other new honeybee species have been recorded for science in the genus *Apis*. These are *Apis andreniformis*, *Apis binghami*, *Apis breviligula*, *Apis koschevnikovi*, *Apis laboriosa*, *Apis nigrocincta*, and *Apis nuluensis*.

These identified species of honeybees nesting one of two different ways, and this nesting behavior determines whether the bees will tolerate being kept inside a manufactured hive. Some of the species make nests consisting of a series of parallel combs, other species nest on just one, single comb. The species that build a series of parallel combs usually nest inside cavities, and this behavior enables them to nest inside manufactured containers and therefore opens up possibilities for the keeping and management of these bees inside hives.

1.2.1 *Apis andreniformis* (Black Dwarf Honeybee)

Apis andreniformis was the fifth honeybee species to be described of the seven known species of *Apis andreniformis* can be distinguished from other *Apis* species by noting their dark black coloration, making them the darkest of their genus. Some distinctions include structural differences in the endophalli, a larger wing venation in *A. andrenifromis*, and a longer basitarsal extension in *A. florea*. Additionally, there are slight color variations between the two species. In *A. andreniformis*, its first two abdominal segments are black and its scutellum is yellow. Another distinguishing factor is the presence of black hairs on the tibia of *A. andreniformis*, which are white in *A. florae*. These are very small-sized species of bees. Their single comb nests are small too; often no larger than 150 - 200cm wide. These bee species construct a single-comb nest, regularly fairly low down in bushes, or in the open, suspended from a branch for *Apis florae* and rock surface for *Apis andreniformis* (Rattana et al., 2007).

1.2.2 *Apis koschevnikovi*

Apis koschevnikovi is of the family Apidae and genus *Apis*. *A. koschevnikovi* is known as one of the 'Red Bees' of Borneo. *A. koschevnikovi* appears together with *A. cerana* and *A. mellifera*, two other cavity-nesting species, in three separate phylogenic clusters without overlapping. The phylogenetic cluster analysis of *A. koschevnikovi* is found directly in contact with a cluster of *A. cerana* and distant from *A. mellifera* (Ruttner et al., 1989). The individual bees are slightly larger than *Apis cerana* found in the same locality, but otherwise the nests of these bees are similar in size and construction (Radloff et al., 2011).

1.2.3 Apis laboriosa

Himalayan honeybee was initially described as a distinct species. Later, it was included in *A. dorsata* as a subspecies (Michael, 1999). Based on the biological species concept, though authors applying a genetic species concept have suggested it should be considered a species. Essentially restricted to the Himalayas, it differs little from the giant honeybee in appearance, but has extensive behavioral adaptations that enable it to nest in the open at high altitudes despite low ambient temperatures. It is the largest living honeybee (Maria & Walter, 2005). Their nests are similar to those of *Apis dorsata*, but *Apis laboriosa* colonies are generally found together in clusters. They are dark in appearance and has long hairy coat and live in high mountains. It constructs a single comb. *Apis laboriosa* is very aggressive.

1.2.4 Apis nigrocincta

A. nigrocincta is a medium sized cavity-nesting species. Colonies are permanent and new colonies are formed by fission (Michener, 2000). Multiple combs are found in dark cavities such as hollows of trunks of live or dead trees, underneath roofs, water jars, and caves. Combs are also built with multiple attachment sites so that the contents of the nest are spread out over several points of contact. These nesting sites are close to the ground, usually 4 – 5m in distance. Multiple combs are built in a pattern in which there is a uniform distance, known as the bee space, between each comb. There are two types of combs in a brood: smaller ones for workers and larger ones for drones, with worker cells averaging around 4.5mm.

Queen cells can be found on the lower edges of the combs, while honey is stored in the upper as well as the outer part of the combs (Radloff et al., 2011). Its nesting behavior is analogous to *Apis cerana* and *Apis koschevnikovi*. Similar to virtually all other Asian honeybee species, *A. nigrocincta* will abandon a nest to start a new one through migration and absconding. Due to the tropical climate of Southeast Asia, conditions for migration and absconding are possible year round, though predation pressure is severe in many areas. Colonies will leave due to disastrous events of nature or situations where abandonment is necessary, or due reduced resources. In addition, one can predict colony migrations due to seasonal declines in pollen or extreme temperature.

1.2.5 Apis nuluensis

Apis nuluensis a new species recently described by Tingek et al. (1996) on the basis of morphological and behavioral characters. *A. nuluensis* shows some extreme characteristics, which separate it from all other cavity-nesting bees, or all *A. cerana* groups. In size measures, it is closest to *A. nigrocincta* from Sulawesi, while wing venation measures are close to the northern *A. cerana* groups. It is a new species of cavity-nesting honeybee. Its nesting behavior is analogous to *Apis cerana* and *Apis koschevnikovi*. The fact that *A. nuluensis* clustered together with *A. cerana*, rather than as a separate group as observed for *A. koschevnikovi*, suggests that *A. nuluensis* diverged from *A. cerana* more recently. Thus, although *A. nuluensis* is morphologically distinct from *A. cerana*, molecular analyses indicate that, for the nuclear and mitochondrial regions examined, few mutations have accumulated since the two species diverged.

1.3 Stingless Bees

Stingless bees belong to the tribe Meliponini, and keeping these bees is known as meliponiculture, analogous to apiculture which refers to the keeping of honeybees. Stingless bees with meliponiculture would provide honey for food and medicine, and enhance pollination of both commercial crops and indigenous plants (Eardley, 2004). Like honey bee (*Apis mellifera*), which provides mostly produced honey, stingless bees have enlarged areas on their back legs for carrying pollen back to the hive. After a foraging expedition, these pollen baskets or corbiculae can be seen stuffed full of bright orange or yellow pollen. Stingless bees also collect nectar, which they store in an extension of their gut called a crop.

Being tropical, stingless bees are active all year round, although they are less active in cooler weather (Alves et al., 2009). They do not sting, but will defend by biting if their nest is disturbed. In addition, a few (in the genus *Oxytrigona*) have mandibular secretions that cause painful blisters. Stingless bees usually nest in hollow trunks, tree branches, underground cavities, or rock crevices, but they have also been encountered in wall cavities, old rubbish bins, and storage drums. Many beekeepers keep the bees in their original log hive or transfer them to a wooden box, as this makes it easier to control the hive. Some beekeepers put them in bamboos, flowerpots, coconut shells, and other recycling containers such as a water jug, a broken guitar, and other safe and closed containers (Villanueva et al., 2005).

The bees store pollen and honey in large, egg-shaped pots made of beeswax (typically) mixed with various types of plant resin; this combination is sometimes referred to as cerumen (which is, incidentally, the medical term for earwax). These pots are often arranged around a central set of horizontal brood combs, where in the larvae are housed. When the young worker bees emerge from their cells, they tend to initially remain inside the hive, performing different jobs. As workers age, they become guards or foragers. Unlike the larvae of honey bees and many social wasps, meliponine larvae are not actively fed by adults (progressive provisioning). Pollen and nectar are placed in a cell, within which an egg is laid, and the cell is sealed until the adult bee emerges after pupation (mass provisioning). At any one time, hives can contain 300 – 80,000 workers, depending on species.

Unlike true honey bees, whose female bees may become workers or queens strictly depending on what kind of food they receive as larvae (queens are fed royal jelly and workers are fed pollen), the caste system in Meliponines is variable, and commonly based simply on the amount of pollen consumed; larger amounts of pollen yield queens in the genus *Melipona*.

> **Honeybee species whose nests consist of multiple or single combs**
>
> Keep in mind that all above species of honeybees' nest in one of two different ways, and this nesting performance determines whether or not the bees can keep inside a manmade (artificial) hives. Some of the species construct nests consisting of a series of parallel combs and other species nest on simply single comb. The species of honey bees that construct a series of parallel combs usually nest within cavities, and this performance enables them to nest within man-made hives, as a result beekeepers should select the best species of bees that can well-matched in manmade hive. The species that build single combs usually nest in the open. They cannot keep in hives and the single comb behavior does not lend itself to beekeeping management practices, although the honey and other products of these species are harvested by some societies.

Table 1.1 Species of honeybees: type of nests

Nests consist of multiple combs (cavity nesting honeybees)	Nests are single combs
Apis cerana	*Apis andreniformis*
Apis koschevnikovi	*Apis binghami*
Apis mellifera	*Apis breviligula*
Apis nigrocincta	*Apis dorsata*
Apis nuluensis	*Apis florea*
	Apis laboriosa

2 RACES OF HONEYBEES AND THEIR BEHAVIOR

In the original home land of *A. mellifera* in Europe, Africa and the Near East, the bees remained under the effect of natural selection for long time. Classification of honey bee races indicate the differences in physical characters between subspecies, their present geographical distribution, the geological evidence pointing to their origins and to the course of their subsequent evolution and distribution. *Apis mellifera* had to adapt itself to a large variety of habitats and climates. This adaptation was achieved by natural selection, producing some two dozen subspecies or races (Ashleigh, 1996).

An essential tool that used for honeybee races discrimination and characterization is morphometric analyses. The common method of morphometric analyses for the characterization and classification of honeybee subspecies are based mainly on measuring honeybee wing characters, which were considered as strong tool (Abou-Shaara and Al-Ghamdi, 2012). Bee differences can be used to advantage by beekeepers, depending on what traits interest them. Thus, it is best for each beekeeper to experience the characteristics of the different bee strains first hand and then form an opinion about which stock best fits his or her situation.

The different races of bees can generally be differentiated in physiological terms. Bees from warmer climates tend to be smaller and lighter in colour than that adapted to the colder regions. Some subspecies are more prone to swarming than others are, some produce large numbers of young queens when swarming, others only a few. Tropical honeybees frequently abscond or migrate, sometimes due to lack of forage through drought or other causes, perhaps as a defence against predators. Heavy predation is also a likely cause of the vigorous defence reaction of some races, for example, the bees of tropical Africa.

Same race have occupied different kinds of habitat, they have formed local strains, which have accommodated themselves to the different conditions. Similarly, honeybees of different races, which have occupied similar habitats, have evolved similar behavioral characters. Even the dance language by which honeybees communicate information about the location of food sources may differ in detail between races as different races may be conditioned to foraging over different distances from the nest.

The behavioral characters of the different races and strains, brood rearing pattern, foraging behavior, clustering, etc., are fixed genetically, so that a colony cannot readily adapt itself when transferred to a different kind of environment. The behavioral patterns, which have evolved in the different races, have ensured the survival of the various subspecies in their native habitats and some of these patterns may be repeated in different races.

2.1 Distinctive Characteristics of Races of Bees

To distinguish one race from the other, different characters have to be used. However, scientists and beekeepers do not use the same criteria. Scientists tend to use the biometric measurements (morphometric & behavioral characters) while beekeepers prefer to use characters like colour and behavioral traits.

> *Colour*: The dorsal abdominal segments vary in colour from light yellow to entirely dark from race to race. However, due to the variability of colour within the same race, its value for characterizing is less important.

> *Length of tongue*: Tongue length is the most important characteristics for identification of races & during selection. In addition, it is important parts influence the productivity of the race.

> *Hair coverage*: The area covered by the bands of hair on the abdomen. The length & colour of hair are important character for race identification. Other distinctive

characteristics that used to distinguish one race from the other are the number of hooks on the wing, width of the metatarsus and the shape and size of wax glands and others.

2.2 European Honeybee Races

The western honeybee, *Apis mellifera* Linnaeus, naturally occurs in Europe, the Middle East, and Africa. This species has been subdivided into at least 20 recognized subspecies (or races), none of which are native to the Americas. However, subspecies of the western honeybee have been spread extensively beyond their natural range due to economic benefits related to pollination and honey production. In the United States, European honeybees represent a complex of several interbreeding European subspecies including *Apis mellifera ligustica* (Italian bees), *Apis mellifera carnica* (Carniolan bee), *Apis mellifera mellifera* (dark bees), *Apis mellifera causcasia* (Russian bee) and *Apis mellifera iberiensis* (Ashley et al., 2013).

2.2.1 *Apis mellifera ligustica* (Italian Bees)

- Usually have bands on their abdomen of brown to yellow color
- Gentle and non-aggressive
- Very good honey producers
- Uses less propolis than some of the darker bees
- Prone to rob and drift
- Robbing behavior may pose problems because it may cause the rapid spread of transmittable diseases
- Colonies are usually large
- Queens lay all through the summer, so a large amount of stores is used for brood rearing
- Extended periods of brood rearing
- Can build colony populations in the spring and maintain them for the entire summer
- May consume surplus honey in the hive if supers are not removed immediately after the honey flow stops
- Swarming instinct is not especially strong
- Less prone to disease than their German counterparts
- Keeps a clean hive
- Quick to get rid of the wax moth

2.2.2 *Apis mellifera mellifera* (Dark Bees)

- Its original homeland is all of Europe, north and west of Alps & central Russia.
- It is the first honeybee brought to the new world.
- At present, these bees are found in some parts of Spain, France, Poland and Russia as a pure race.
- Exteriorly large (big in size)

- Has broad & large abdomen
- Black in colour but same times with yellow spot on abdomen
- Short tongues: 5.7 – 7.4mm
- Narrow wings
- Long body hair
- Nervous & aggressive
- Low in honey production
- Low tendency to replace its colony
- Weak disposition to swarming
- Dark bees is much inferior to the long tongued races in collection of nectar from clover.
- Susceptible to brood diseases and wax moth
- Strong tendency to sting

2.2.3 *Apis mellifera carnica* (Carniolan Bees)
- It is darker brown to black and a very gentle race of bees (Fig. 3).
- Probably the best wintering bees and little use of propolis
- Builds up very rapidly in the spring to take advantage of blooms
- Summer brood rearing depends on pollen and nectar flow
- Usually not inclined to rob and these bees tend to swarm more
- Not as productive as Italian bees
- Extremely docile and can be worked with little smoke and clothing
- Much less prone to robbing other colonies of honey
- Lowering disease transmission among colonies
- Very good builders of wax combs
- They fly in slightly cooler weather.
- In theory, better in northern climates
- Frugal for the winter and winters in small clusters
- Shuts down brood rearing when there are dearth

2.2.4 *Apis mellifera caucasica* (Russian Bees)
- Tends to rear brood only during times of nectar and pollen flows
- Brood rearing and colony populations tend to fluctuate with the environment
- Good housecleaning behaviors, resulting in resistance not only to Varroa but also to the tracheal mite
- Tends to have queen cells present in their colonies almost all the time
- Performs better when not in the presence of other bee strains
- A bit defensive, but in odd ways
- Traits are not well fixed
- Swarminess and productivity are a bit more unpredictable.
- Frugality is similar to the Carniolan bees.

2.3 African Honeybee Races

These bees are very prone to sting and swarm. They can be managed but with greater difficulty than familiar strains of honeybees. The bees are very dangerous because they are easily disturbed and will attack in great numbers. They also swarm and abscond frequently, so they can be quite difficult for the beekeeper to manage. The greatest danger is to people and animals that cannot run away or get indoors quickly. With the right protective equipment, and by locating hives away from houses and farm animals, we can manage the bees effectively. When you chose your bees, it is good to know that stock sold is rarely pure (Thomas, 2013).

There is high variability between the African bees because of extreme difference in very vast area of the continent and due to adaptation to different tropical conditions. Compared to European honeybees, African races have small body size and have strong tendency to swarm, abscond and migrate. These traits have developed and maintained by constant selection pressures from environmental factors. African honeybee races have short developmental time & rapidly reproduce. The periodical losses from different cause would be balanced by very high prolific nature and swarming ability in favourable seasons. This may be an adaptation to the losses due to many factors (predators, pests etc.).

African honeybees (*Apis mellifera scutellata*) are native to sub-Saharan Africa and introduced in the Americas to improve honey production in the tropics. These African honey bees were accidentally released and began to interbreed with European honey bees (*Apis mellifera ligustica*), the most common subspecies used for pollination and honey production in the United States. As a result, the hybrid offspring are called Africanized because of their shared characteristics (Erin et al., 2010). Africanized honey bees are about the same size, shape and color as European honeybees, and a genetic analysis must be used to distinguish one from the other. Africanized honeybees produce more drones and their colonies grow faster than European honeybees. Adults tend to swarm more frequently and apt to completely abandon the hive if disturbed. Africanized honeybees are more likely to nest in small locations, like water meter boxes, cement blocks, old tires and grills. Colonies are more likely to nest underground and migrate for food. The colony dedicates more members to 'guard' the nest and deploys greater numbers for defense when threatened.

2.3.1 *Apis mellifera intermissa*
- It occupies the countries of North Africa, from Morocco to Libya.
- It has very long proboscis.
- It is small.
- Very dark bee
- Bad temper
- High tendency to swarm

- Excellent in honey production in mostly extreme climatic condition of North Africa
- It constructs about 100 queen cells during swarming.
- It has the ability to resist drought.

2.3.2 *Apis mellifera saharansis* (Saharan Bees)
- Found in fertile parts of Sahara Desert, along south edge of mountain Morocco
- It tolerates extreme change of temperature range from 8 – 50°C.
- Small in size than *Apis mellifera intermissa*
- Have moderate tendency to swarm
- Are docile, i.e. not aggressive
- Not effective at defending its nest

2.3.3 *Apis mellifera lamarckii* (Egyptian Bees)
- Medium size with thick hair
- Less aggressive
- Poor in honey production
- It is restricted to the Nile valley, North of Nubian.

2.3.4 *Apis mellifera jemenitica*
- Very small in size
- It is called honeybee of hot-arid zone of Eastern Africa & Arabia.
- Yellow in color
- Have short morphometric characters like hair, legs, wings & proboscis
- Exhibit variable behavior & morphometric characters
- Exist in high temperature range (27 – 31°C) and low rain fall range of 30 – 300mm

2.3.5 *Apis mellifera scutellata*
- It is native to Tanzania, Burundi, Kenya & Ethiopia (Fig. 4).
- Found in central and eastern equatorial Africa, South of Sahara
- In savannah land of Africa from semi desert to tropical rain forest
- Small in size, very aggressive and their management is difficult
- Has short tongue & slender body
- Brood rearing is rapid. The queen is very prolific. The colonies grow much more rapidly than those of European bees and forage intensively.
- Very high swarming tendency
- Worker bees mature in 19 – 20 days as compared to 21 days for European bees.

2.3.6 *Apis mellifera litorea*
- It is found in the humid coastal regions of East Africa, like Tanzania.
- Small body size
- Has longer tongue
- It is yellow strip bee.

2.3.7 *Apis mellifera monitcola*
- It is found in the mountain regions of East Africa, i.e. Tanzania, Ethiopia & Kenya

Mountain in altitudes of more than 2,400m.
- It is large.
- Good honey producer
- Relatively gentle
- Its hair is longer than those any other African bees.

2.3.8 Apis mellifera adansonii
- In the earlier description, this name was uniformly used for all races of bees of South of Sahara.
- *A. mellifera adansonii* sometimes called as African bees.
- Its abdomen is remarkably broad.
- Its hair is externally short.
- It has tendency to form migratory swarm.

2.3.9 Apis mellifera unicola
- It is native to Madagascar.
- Has short tongue & legs
- Easy to handle & gentle
- Is uniformly black honeybee
- Found between 1,000 - 2,000m above sea level
- Exhibit variability in behavioral characters

2.3.10 Apis mellifera capensis (Cape Town Bees)
- It lives in a very restricted area in Southwest of South Africa (region of Cape Town).
- Bees are darker.
- With short tongue
- The most remarkable feature of these bees is in queenless colony the ability of workers to lay eggs without mating to produce female brood from which queen may be reared.
- The laying workers have spermatheca but not filled with sperm.
- In the queenless colony, the workers soon start laying eggs, in which high percentage of them develop without fertilization into females.
- She is attractive by other bees.
- She also secrets queen substances.
- It also suppressed the development of ovaries of the other worker bees.

2.4 Differences between African and European Honey Bees

African honeybees and European honeybees are the same species (*Apis mellifera*), but the two are classified as different sub-species. European honeybees have been selected by beekeepers for their strong honey production and storage behavior, their reduced regular swarming (colony splitting) tendencies, and their gentleness. The African honeybee is considerably more defensive than its European cousin is. Consequently, it is important to understand key differences between the defensive African bee and the docile European honeybee (O'Malley et al., 2008). The venom of the African honeybee is no more potent than that of the European honeybee. For a fatality to occur from venom toxicity, it normally would take about 10 stings per pound of body weight, from either an African or a European honeybee. The main difference between the European and African honeybee is the defense response: an African honeybee colony, if disturbed, will send more guard bees to sting, and will pursue for a longer distance and stay agitated for a longer period, than will a European honeybee colony.

The other difference found between African and European honeybees were a few behavioral traits in the worker bees that were all related to the workers' food preference. It was found that *Apis mellifera scutellata* workers focused on pollen processing behaviors while European workers focused on nectar processing behaviors (Fewell et al., 2002). African bees were also more likely to store pollen while European bees stored honey. The study found that worker food preferences determined whether the colony maintained a certain reproductive rate. For example, having fewer or relatively older workers who prefer nectar means that the colony will not have the resources available to feed efficiently new broods.

Table 1.2 Differences between African and European honey bees

African honey bees	European honey bees
May send out 10 – 20 guard bees in response to disturbances up to 20 feet away	May send out several hundred guard bees in response to disturbances up to 120 feet away
Once agitated, will usually become calm within 10 – 15 minutes	Once agitated, may remain defensive for hours or until the sun sets
Disturbing a colony may result in 10 – 20 stings	Disturbing a colony may result in 100 – 1,000 stings
Swarm 1 or 2 times per year	Can swarm 10 or more times a year
Swarms are larger and need larger volume in which to nest	Swarms contain fewer individuals, and therefore a much smaller nest cavity is needed
Rarely abscond (or completely abandon nest) from nesting location	Abscond often and relocate to more suitable nesting locations
Nests in large cavities, around 10 gallons in size	Nests in smaller cavities, 1 to 5 gallons in size
Typically nest in dry, above ground cavities	Will nest in underground cavities
Nests in protected locations, rarely exposing the nest	Will nest in exposed locations (e.g., hanging from a tree branch)
Due to larger colony size, nests are often easier to detect	Due to smaller colony size, nests often go undetected until disturbed

Source: O'Malley et al., 2008.

2.5 Ethiopian Honeybee Races

Being part of Africa, Ethiopian honeybee races are very similar to that of African races.

IDENTIFY HONEYBEE SPECIES, RACES AND THEIR BEHAVIOR

According to study done on morph clusters of geographical races of Ethiopian honeybees by Amssalu et al. (2004), the following five honeybee races have been reported to exist in the country. Those five honeybee races occupying ecologically different areas: *Apis mellifera jemenitica* in the northwest and eastern arid and semi-arid lowlands; *A. mellifera scutellata* in the west, south and southwest humid midlands; *A. mellifera bandasii*, in the central moist highlands; *A. mellifera monticola* from the northern mountainous highlands; and *A. mellifera woyi-gambell* in south western semi-arid to sub-humid lowland parts of the country. Moreover, some areas with high inter and intracolonial variances were noted, suggesting introgression among these defined honeybee populations.

In recent studies by Marina et al. (2011), honeybee's endemic to the volcanic dome system of Ethiopia are described as a new subspecies, *Apis mellifera simensis*, based on morphometrical analyses. Principal component and discriminant analyses show that the Ethiopian bees are clearly distinct and statistically separable from honeybees belonging to neighbouring subspecies in eastern Africa.

2.5.1 Apis mellifera monticola
- The biggest and darkest of all other races found in the country
- Found to exist in the northern high mountainous part of the country
- Has low tendency for reproductive swarming and migration
- Less aggressive than other races
- Has longest body hair than other races

2.5.2 Apis mellifera bandansii
- The largest honeybees next to *A. m. monticola*
- Found in central highlands of the country
- Dark in colour, but has few yellow members
- Has longest body hair next to *A. m. monticola*
- Has high tendency for reproductive swarming
- Has less migration tendency than *A. mellifera jemenitica*
- Is less aggressive than *A. mellifera jemenitica*
- Give better honey yield than *A. mellifera jemenitica*

2.5.3 Apis mellifera scutelltata
- Occupy the wet tropical forest lands
- It is darker than *A. mellifera jemenitica* & *A. mellifera woyi-gambella*.
- Its population comprises some yellow honeybees.
- Has higher tendency for migration
- It exhibits aggressive to highly aggressive behavior.
- Give better yield than *A. mellifera jemenitica*

2.5.4 Apis mellifera jemenitica
- Is the yellowest honeybee but also consists black members?
- Smaller than *A. m. bandansii*, *A. m. monticola* & *A. m. scutellate*

- Has less tendency for reproductive swarming
- Has high migration tendency
- Is aggressive than other races

2.5.5 Apis mellifera woyi-gambella
- Found in the extreme western and southern semi-arid to sub moist low lands
- Found only in Ethiopia
- It is the smallest of all honeybee races in the world.
- It has the shortest hair cover.
- It is predominantly yellow in colour, but also comprise black members.
- Has less tendency for reproductive swarming
- Has intermediate migration behavior
- It is aggressive to highly aggressive in behavior.

>>> SELF-CHECK QUESTIONS

Part 1. Choose the correct answer from the given alternative and encircle the answer.

1. Which of the following honeybee species build series of parallel combs?
   A. *Apis cerana*   B. *Apis koschevnikovi*   C. *Apis mellifera*   D. All
2. Which of the following honeybee species build single combs?
   A. *Apis nigrocincta*   B. *Apis nuluensis*
   C. *Apis dorsata*   D. *Apis cerana*
3. Which of the following honeybee species constitutes integral part of modern agricultural system?
   A. *Apis mellifera*   B. *Apis breviligula*
   C. *Apis koschevnikovi*   D. *Apis laboriosa*
4. Which of the following indicate the advantage of Italian Bees?
   A. Gentle and non-aggressive   B. Very good honey producer
   C. keeps a clean hive   D. All
5. _____ is honeybee race that found only in Ethiopia.
   A. *Apis mellifera scutelltata*   B. *Apis mellifera jemenitica*
   C. *Apis mellifera woyi-gambella* D. *Apis mellifera bandansii*
6. Among honeybees that found in Ethiopia, which one is less aggressive than other races?
   A. *Apis mellifera monticola*   B. *Apis mellifera bandansii*
   C. *Apis mellifera jemenitica*   D. *Apis mellifera scutelltata*
7. _____ is the biggest and darkest of all other honeybee races found in Ethiopia.
   A. *Apis mellifera monticola*   B. *Apis mellifera woyi-gambella*
   C. *Apis mellifera jemenitica*   D. *Apis mellifera scutelltata*

8. Which one of the following sentence indicate the behavior of *Apis mellifera intermissa*?
   A. It has very long proboscis.
   B. It is small in size.
   C. Very dark bee
   D. All

9. Which one of the following sentence indicate the behavior of *Apis mellifera scutellata*?
   A. large in size
   B. Very aggressive and their management is difficult
   C. Has long tongue
   D. Less aggressive

10. Which one of the following sentence indicate the behavior of *Apis mellifera lamarckii* (Egyptian bee)?
    A. Large size with thick hair
    B. More aggressive
    C. Poor in honey production
    D. High honey yield

Part 2. Match the correct answer from column B to column A and write the letter you choice on the space provided.

A	B
_____ 1. Oriental honeybee	A. *Apis mellifera*
_____ 2. *Apis florea*	B. Dwarf honeybee
_____ 3. Most wide spread species	C. Black dwarf honeybee
_____ 4. *Apis dorsata*	D. Giant honeybee
_____ 5. *Apis mellifera mellifera*	E. Dark bees
_____ 6. *Apis mellifera ligustica*	F. Italian bees
_____ 7. *Apis mellifera caucasica*	G. Russian bees
_____ 8. *Apis mellifera capensis*	H. Cape Town bees
_____ 9. *Apis mellifera lamarckii*	I. egyptian bee
_____ 10. *Apis andreniformis*	J. *Apis cerana*
	K. Saharan bees
	L. *Apis laboriosa*

Part 3. Write the correct answer accordingly.
1. Write commonly recognized honeybee species.
2. Describe the unique characters of commonly recognized honeybee species.
3. Write common honeybee races.
4. Write the advantage and disadvantages of common honeybee races.

5. Write the importance of identifying the unique character of honeybee species and races before introducing to a given apiaries.

>>> REFERENCES

Abou-Shaara H F, Al-Ghamdi A, 2012. Studies on wings symmetry and honey beeraces discrimination by using standard andgeometric morphometrics [J]. Biotechnology in Animal Husbandry, 28 (3): 575-584.

Alves D A, Imperatriz-Fonseca V L, Santos-Filho P S, 2009. Production of workers, queens and males in *Plebeia remota* colonies (Hymenoptera, Apidae, Meliponini), a stingless bee with reproductive diapause [J]. Genetics and Molecular Research, 8 (2): 672-683.

Amssalu B, Nuru A, Radloff S E, et al., 2004. Multivariate morphometric analysis of honeybees (*Apis mellifera*) in the Ethiopian region [J]. Apidologie, 35: 71-81.

Arias M C, Sheppard W S, 2005. Phylogenetic relationships of honey bees (Hymenoptera: Apinae: Apini) inferred from nuclear and mitochondrial DNA sequence data [J]. Mol Phylogenet Evol, 37 (1): 25-35.

Danforth B N, Sipes S, Fang J, et al, 2006. The history of early bee diversification based on five genes plus morphology [J]. PNAS, 103 (41): 15118-15123.

Eardley C D, 2004. Taxonomic revision of the African stingless bees (Apoidea: Apidae: Apinae: Meliponini) [J]. African Plant Protection, 10 (2): 63-96.

Engel M S, 1999. The taxonomy of recent and fossil honey bees (Hymenoptera: Apidae: *Apis*) [J]. Journal of Hymenoptera Research, 8: 165-196.

Fewell J H, Bertram S M, 2002. Evidence for genetic variation in worker task performance by African and European honeybees [J]. Behavioural Ecology and Sociobiology, 52: 318-325.

Jones J C, Nanork P, Oldroyd B P, 2007. The Role of Genetic Diversityin Nest Cooling in a Wild Honey Bee, Apis Florea [J]. Journal of Comparative Physiology, 193: 2.

Meixner M D, Leta M A, Koeniger N, et al., 2011. The honey bees of Ethiopia represent a new subspecies of *Apis mellifera—Apis mellifera simensis* n. ssp. [J]. Apidologie, 42: 425-437.

Paar J, 2006. Genetic Structure of an *Apis dorsata* Population: The Significance of Migration and Colony Aggregation [J]. Journal of Heredity, A12: 119-126.

Radloff S E, Hepburn H R, Engel M S, 2011. Honeybees of Asia [M]. Berlin: Springer Science & Business Media.

Seeley, Thomas D. 2009. The wisdom of the hive: the social physiology of honeybee colonies [M]. Harvard University Press.

Tingek S, Koeniger N, Koeniger G, 1996. Description of a new cavity dwelling species of *Apis* (*Apis nuluensis*) from Sabah, Borneo with notes on its occurrence and reproductive biology (Hymenoptera, Apoidea, Apini) [J]. Senckenbergiana Biol, 76: 115-119.

Villanueva-G R, Roubik D W, Colli-Ucan W, 2005. Extinction of *Melipona beecheii* and traditional beekeeping in the Yucatan peninsula [J]. Bee World, 86: 35-41.

MODULE 2:
RECOGNIZE HONEY BEE COLONY AND ITS ORGANIZATION

>>> INTRODUCTION

When you first look into a hive and see thousands of bees apparently moving around at random and flying off the comb in all directions, they appear confused. But it isn't. All this movement has a purpose and, within a short time experience in beekeeping, you will begin to understand for what purpose it is, and that is a highly organized society going about its business. You will also notice when things aren't going right in the colony and, with more experience, you will be able to look at each comb and, almost instantly, will be able to picture clearly in your mind the state of the colony. Is it healthy? Is there a queen? Is the queen laying well? Are the bees building up in numbers as you would expect? Do they need feeding? It is like reading a book with clearly drawn diagrams. First, however, you should gain an understanding about the bees and what they need for their survival. Only then can the beekeeper work with bees, adapting his or her requirements to theirs (David, 2008). This module therefore deals about honey bee colony and its social organization. At the end of this module you will be able to: identify the members (castes) of bee colony and their duties and responsibilities, understand the developmental stages (metamorphosis) of bees and understand the communication systems of bees.

1 HONEY BEE CASTES AND THEIR DUTIES IN THE COLONY

Honey bees are social insects, which mean that they live together in large, well-organized family groups. A single honey-bee cannot live for very long on its own (Clarence, 2004). When you first look into a hive and see thousands of bees apparently moving around at random and flying off the comb in all directions, you may think that all those insects look exactly the same, actually they are three different kinds of bees: workers, drones, and a queen (Fig. 5). Each has its own characteristics, roles, and responsibilities (Howland, 2009; David, 2008).

1.1 Queen Bee

Queen bee is a fully developed female bee capable of laying eggs to assure the sustainability and growth of the colony. She is the mother of the rest members (workers & drones) of the colony (Patricia and David, 2007). Each colony has only one queen, except during and a varying period following swarming. She is easily distinguished from other members of the colony. Her body is normally much longer than either the drone's or worker's, especially during the egg-laying period. Her wasp like slender abdomen usually without color bands distinguishes her from both workers and drones. Her wings cover only about two-thirds of the abdomen. A queen's thorax is slightly larger than that of a worker. Viewed from the front, her head is round (Diana and Alphonse, 1998; Clarence, 2004; Segeren, 2004).

The queen bee is the heart and soul of the colony. She is the reason for nearly everything the rest of the colony does. The queen is the only bee without which the rest of the colony members cannot survive. The other bees pay close attention to the queen, tending to her every need.

Like a regal celebrity, she's always surrounded by a flock of attendants as she moves about the hive (Fig. 6). These attendants are vital, because the queen is totally incapable of tending to her own basic needs. She can neither feed nor groom herself. She can't even leave the hive to relieve herself, so her doting attendants (the queen's court) take care of her basic needs while she tirelessly goes from cell to cell doing what she does best (Howland, 2009).

Because queen is the only sexually developed (has fully developed ovaries) female, her primary function is egg laying. She is, in fact, an egg-laying machine, capable of producing more than 1,500 eggs a day at 30 – second intervals. That many eggs are more than her body weight (Howland, 2009). But, the number of eggs the queen lays depends on the amount of food she receives and the size of the worker force capable of preparing beeswax cells for her eggs and caring for the larva that will hatch from the eggs in 3 days. She can produce both fertilized and unfertilized eggs. If her egg is laid in a larger drone-sized cell, she normally does not release sperm (unfertilized egg), and the resulting individual becomes a drone. If she lay eggs in the worker (smaller sized comb cells) cells and queen cells (a special cell, shaped like a peanut, which is usually suspended vertically from the lower part of the comb), they are fertilized and develop in to adult worker and queen bees respectively.

The second major function of a queen is producing pheromones that serve as a social 'glue' unifying and helping to give individual identity to a bee colony (Clarence, 2004). The queen produces a number of different pheromones in her mandibular (jaw) glands that attract workers to her and stimulate brood rearing, foraging, comb building, and other activities. Also referred to as *queen substances*, these pheromones play an important role in controlling the behavior of the colony. Queen substances inhibit the worker bees from making a new queen and prevent the development of the worker bees' ovaries, thus ensuring that the queen is the only egg-laying female in the hive. They act as a chemical communication that 'all is well—the queen is in residence and at work.' As a queen ages, these pheromones diminish, and, when that happens, the colony knows that it's time to supersede her with a new, young queen.

Pheromones are essential in controlling the well-being of the colony. This queen substance makes its way around the hive like a bucket brigade. The queen's attendants pick up the scent from the queen and transfer it by contact with neighboring bees. They in turn pass the scent onto others, and so it distributes throughout the colony. So effective is this relay, that if the queen were removed from the hive, the entire colony would be aware of her loss within hours. When the workers sense the lack of a queen, they become listless, and their drive to be productive is lost. Without leadership, they nearly lose their reason for being. First, they're unhappy and mope around, but then it dawns on them 'let's make a new queen' (Howland, 2009).

1.2 Worker Bees

Worker bees are the smallest bodied adults and constitute the majority of bees occupying the colony. Although they are females, they lack the fully developed reproductive organs. They are infertile (sexually undeveloped) females and under normal hive conditions do not lay eggs, although if the colony is without a queen, a few workers may begin to lay eggs which develop in to drone bees (Clarence, 2004).

Workers look different (Fig. 5) than the queen (Howland, 2009). Viewed from the front the worker has a triangular shape. Their abdomens are shorter, and the tips of their wings in the rest position cover the end of their abdomen. One segment of their hind legs bears two rows of long hairs between which the pollen is carried. The pollen that covers the bee's body after her visit to a flower is stored and transported in these baskets (Segeren, 2004).

From the moment a worker bee emerges from her cell she has many and varied tasks clearly cut out for her. As she ages, she performs more and more complex and demanding tasks. Although these various duties usually follow a set pattern and timeline, they sometimes overlap. A worker bee may change occupations sometimes within minutes, if there is an urgent need within the colony for a particular task. They represent teamwork and empowerment at their best. Initially, a worker's responsibilities include various tasks within the hive. At this stage of development, worker bees are referred to as house bees. As they get older, their duties involve work outside of the hive as field bees.

Responsibilities of worker bees

In the following paragraphs, the various responsibilities of worker bees explained by

Diana and Alphonse (1998) and Howland (2009) are highlighted.

1) Cleaning and maintaining the hygiene of the nest

On emerging from her cell as an adult bee, the worker begins work by cleaning out brood cells from which she just emerged. This and other empty cells are cleaned and polished and left immaculate to receive new eggs and to store nectar. Cell preparation is accomplished by very young workers, only a few hours old. These young bees remove nearby cocoon remains and larval feces from brood cells.

During the first couple weeks of her life, the worker bee removes any bees that have died and disposes of the corpses as far from the hive as possible. Similarly, diseased or dead brood is quickly removed before becoming a health threat to the colony. Should a larger invader (such as a mouse) be stung to death within the hive, the workers have an effective way of dealing with that situation.

Obviously, a dead mouse is too big for the bees to carry off. So, the workers completely encase the corpse with *propolis* (a brown sticky resin collected from trees, and sometimes referred to as *bee glue*). This hygienic behavior is a genetic trait, one that is desirable for the beekeepers to perpetuate. For instance, the continual quick removal of dead brood from the cells and the hive is considered hygienic behavior. Colonies whose workers demonstrate good hygiene are more likely to be free from some diseases.

2) Caring for larvae

The young worker bees tend to their 'baby sisters' by feeding and caring for the developing larvae. These bees are often referred to as nurse bees. On average, nurse bees check a single larva 1,300 times a day. They feed the larvae a mixture of pollen and honey, and royal jelly—rich in protein and vitamins—produced from the hypopharyngeal gland in the worker bee's head and mandibular gland.

The number of days spent tending brood depends upon the quantity of brood in the hive, and the urgency of other competing tasks. You can see these bees sticking their heads into cells for a few seconds to determine how much food is available and to feed the larvae as needed. The brood food is placed near the cell bottom, close to the larva's mouth.

3) Caring for queen

In addition, these young nurse bees also feed, clean and groom the queen. Because her royal highness is unable to tend to her most basic needs by herself, some of the workers do these tasks for her. They form a retinue or circle around the queen. Each attendant stays only from about one to three minutes and then departs. Another attendant takes her place, so the queen is always surrounded, except if she is moving quickly across the comb. In the process these tactile activities, the attendants collect and distribute queen pheromones. When departing attendants contact other bees in the hive, they pass on the queen's scent adhering to their bodies especially on the antennae. The bees feed the queen royal jelly directly into her mouth. They also groom her and even remove her excrement from the hive. These royal attendants also coax the queen to continuously lay eggs as she moves about the hive.

4) Unloading the pollen and nectar from field bees

These young worker bees' another task is receiving nectar from foraging field bees that are returning to the hive. The house bees deposit this nectar into cells earmarked for this purpose. They add an enzyme to the nectar and set about fanning the cells to evaporate the water content and turn the nectar into ripened honey. The workers similarly take pollen from returning field bees and pack down with their heads into cells moisten it with honey and saliva (which contains enzymes), turning the mixture in to *bee bread*. When a cell is filled with bee bread, it is covered with a layer of honey. Both the ripened honey and the pollen (bee bread) are food for the colony.

5) Ventilating the nest

Worker bees also take a turn at controlling the temperature and humidity of the hive. During warm weather and during the honey flow season, you'll see groups of bees lined up at one side of the entrance, facing the hive. They often be seen fanning their wings on the extended deck of the bottom board with their heads facing toward the hive entrance. In this position, warm air is pulled out and fresh air is drawn into the hive. Additional fanners are in position within the hives. This relay of fresh air helps maintain a constant temperature (93 to 95 °F) for developing brood. The fanning also hastens the evaporation of excess moisture from the curing honey.

6) Comb construction

Worker bees that are about 12 days old are mature enough to begin producing beeswax. These white flakes of wax are secreted from wax glands on the underside of the worker bee's abdomen. Worker bees chew these small flakes of wax and work them to form the comb. The honeycomb is the inner house of honey bees. It is where young bees are raised and where the hive's food is stored. They also use it for capping of ripened honey and cells containing developing pupae.

7) Guarding

The last task of a house bee before she ventures out is that of guarding the hive. They defend their hive by flying at and often stinging an intruder. At this stage of maturity, their sting glands have developed to contain an authoritative amount of venom. You can easily spot the guard bees at the hive's entrance. They are poised and alert, checking each bee that returns to the hive for a familiar scent. Only family members can pass. Strange bees, wasps, hornets, and others intent on robbing the hives vast stores of honey are bravely driven off.

8) Foraging

With her life half over, the worker bee now ventures outside of the hive and joins the ranks of field bees. You'll see them taking their first *orientation flights*. The bees face the hive and dart up, down, and all around the entrance. They're imprinting the look and location of their home before beginning to circle the hive and progressively widening those circles, learning landmarks that ultimately will guide them back home. At this point, worker bees are foraging for pollen, nectar, water, and propolis (resin collected from trees).

They forage a 2 to 3 mile (4 to 5 kilometer) radius from the hive in search of food. They're ready and willing to travel. Foraging is the toughest time for the worker bee. It's difficult and dangerous work, and it takes its toll. They can get chilled as dusk approaches and die before they can return to the hive. Sometimes they become a tasty meal for a bird or other insect. They've grown darker in color, and their wings are torn and tattered. This is how the worker bee's life draws to a close working diligently right until the end (Fig. 7).

1.3 Drone Bees

The drones are the male bees in the colony. Usually several hundred to several thousand drones are present in a colony during the active foraging season. They are larger and heavier than the workers, but not as long as the queen. It is easy to identify a drone by its large compound eyes that come together at the top of the head (Clarence, 2004). Because drones are larger, beginners often mistake them for queens. But his shape is in fact more like a barrel (the queen's shape is thinner, more delicate and tapered). The drone's eyes are huge and seem to cover his entire head. He doesn't forage for food from flowers—he has no pollen baskets. He doesn't help with the building of comb—he has no wax-producing glands, nor can he help defend the hive—he has no stinger and can be handled by the beekeeper with absolute confidence (Howland, 2009). The main task of drone is to mate with a young queen. If there is a shortage of food in the colony the drones are no longer fed and after a while they are dragged out of the hive (Segeren, 2004).

2 DEVELOPMENT STAGES OF BEES

Each type of bee begins life as a small egg laid by the queen (Fig. 2.1) in the base of a wax cell in the comb (David, 2008). Like butterflies, honey bees develop in four distinct phases: egg, larva, pupa, and adult. The detailed development stages are shown in Fig. 2.2.

The total development time varies a bit among the three castes of bees, but the basic miraculous process is the same: 24 days for drones, 21 days for worker bees, and 16 days for queens (Howland, 2009).

2.1 Egg

The metamorphosis begins when the queen lays an egg. Honey bee eggs are normally laid one per cell by the queen. Each egg is attached to the cell bottom and looks like a tiny grain of rice (Fig. 2.1). When first laid, the egg stands straight up on end. However, during the 3-day development period, the egg begins to bend over. On the third day, the egg develops into a tiny grub and the larval stage begins (Clarence, 2004).

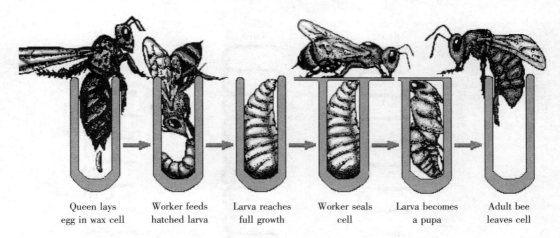

Fig. 2.1 Diagram of the life cycle of the honeybee
Source: Ministry of Agriculture, Animal Industry and Fisheries of Uganda, 2012.

You should know how to spot eggs, because that is one of the most basic and important skills you need to develop as a beekeeper. It isn't an easy task, because the eggs are mighty tiny (only about 1.7 millimeters long). But finding eggs is one of the surest ways to confirm that your queen is alive and well. It's a skill you'll use just about every time you visit your hive. The queen lays a single egg in each cell that has been cleaned and prepared by the workers to raise new brood.

The queen lays the egg in an upright position (standing on end) at the bottom of a cell. That is why they are so hard to see. When you look straight down into the cell, you are looking at the miniscule diameter of the egg, which is only 0.4 of a millimeter wide.

Eggs are much easier to spot on a bright sunny day. Hold the comb at a slight angle, and with the sun behind you and shining over your shoulder, illuminate the deep recesses of the cell. The eggs are translucent white, and resemble a miniature grain of rice (Howland, 2009).

If the queen chooses a standard worker-size cell, she lays a fertilized egg into the cell. That egg develops into a worker bee (female). The fertilized eggs can also develop in to queen bee (the detail is dealt in module 10). But if she chooses a wider drone-size cell, the queen lays unfertilized egg. That egg develops into a drone bee (male). The workers that build the cells are the ones that regulate the ratio of female worker bees to male drone bees. They do this by building smaller size cells for female worker bees, and larger size cells for male drone bees (Howland, 2009).

2.2 Larva

Three days after the queen lays the egg, it hatches into a *larva* (the plural is *larvae*). Healthy larvae are snowy white with a glistening appearance. They are curled in a 'C' shape on the bottom of the cell. Tiny at first, the larvae grow quickly, shedding their skin five

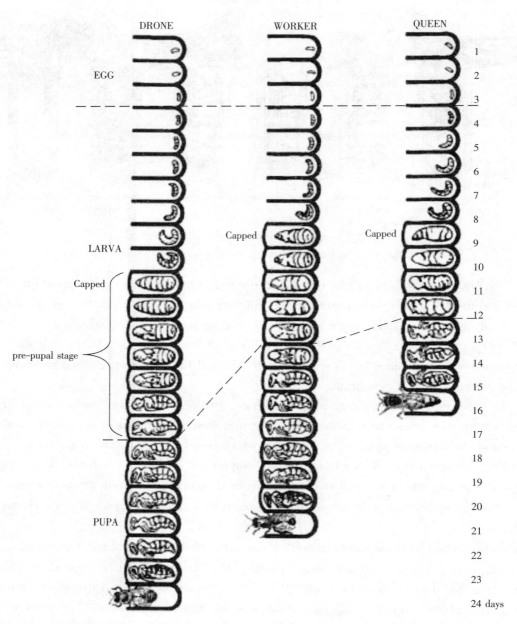

Fig. 2.2 Diagram of the developmental stages of worker, queen & drone bees
Source: Howland, 2009.

times. These helpless little creatures have voracious appetites, consuming 1,300 meals a day. The nurse bees first feed the larvae royal jelly, and later they're weaned to a mixture of honey and pollen (bee bread). Within just five days, they are 1,570 times larger than their original size. At this time, the worker bees seal the larvae in the cell with a porous capping of tan beeswax.

The period just after the cell is capped is called the pre-pupal stage. Once sealed in, the larvae spin a cocoon around their bodies. During this stage the larva is still grub-like in appearance but stretches itself out lengthwise in the cell and spins a thin silken cocoon. Larvae remain pearly white, plump, and glistening during the pre-pupal stage (Clarence, 2004; Howland, 2009).

2.3 Pupa

Within the individual cells capped with beeswax, the pre-pupae begin to change from their larval form to adult bees. The larva is now changed to *pupa* (the plural is *pupae*). Here's where things really begin to happen (Fig. 8). Healthy pupae remain white and glistening during the initial stages of development, even though their bodies begin to take on adult forms. The eyes, legs, and wings take shape. Compound eyes are the first feature that begin to take on color; changing from white to brownish-purple. Soon after this, the rest of the body begins to take on the color of an adult bee. Finally, the fine hairs that cover the bee's body develop. Of course, the transformations now taking place are hidden from sight under the wax cappings (Clarence, 2004; Howland, 2009). After 16, 21 and 24 days, the eggs laid by the queen develop in to adult queen, worker and drone bee respectively. At the day of their emergence from the cell, they chew their way through the wax capping to join their sisters and brothers. Fig. 2.2 shows the entire life cycle of the three castes of honey bee from start to finish.

3 COMMUNICATION SYSTEM OF HONEY BEES

The entire activity of honeybees depends on communication. Bees can transfer information & receive information. Like human being, honey bees utilize five sense organs throughout their daily lives. However, they have two additional communication aids at their disposal. These additional aids of their communication are dance (movement of the body) and chemicals (pheromones).

3.1 Pheromones

Pheromones are chemical scents that animals produce to trigger behavioral responses from the other members of the same species. Honey-bee pheromones provide the 'glue' that holds the colony together. The three castes of bees produce various pheromones at various times to stimulate specific behaviors.

The study of pheromones is a topic can be entire book (Howland, 2009), so here are just a few basic facts about the ways pheromones help bees communicate:
- Certain queen pheromones (known as *queen substance*,) let the entire colony know that the queen is in residence.
- Outside of the hive, the queen pheromones act as a sex attractant to potential suitors

(male drone bees).
- Queen pheromones stimulate many worker bee activities, such as comb building, brood rearing, foraging, and food storage.
- The worker bees at the hive's entrance produce pheromones that help guide foraging bees back to their hive. The Nassanoff gland at the tip of the worker bee's abdomen is responsible for this alluring scent.
- Worker bees produce alarm pheromones that can trigger sudden and decisive aggression from the colony.
- The colony's brood (developing bee larvae and pupae) secretes special pheromones that help worker bees recognize the brood's gender, stage of development, and feeding needs.

3.2 Dances

The gathering of pollen, nectar and propolis for feeding larvae and for storage requires a high degree of cooperation and communication among the members of the colony. Communications among the bees increases the efficiency of food gathering activities by directing bees to known water and food sources (Diana and Alphonse, 1998). The most famous and fascinating 'language' of the honey bee is communicated through a series of dances. The worker bees dance on the comb using precise patterns. Depending upon the style of dance, worker bees share a variety of information with their sisters. They are able to obtain remarkably accurate information about the location and type of food the foraging bees have discovered (Howland, 2009).

A worker bee orients herself according to the following (Diana and Alphonse, 1998) varies external stimuli as she comes from and goes to food collecting locations:
- The sun's position and polarized light
- Landmarks, both horizontal and vertical
- Ultraviolet light, which enables her to see the sun on cloudy days

Two common types of dances are the so-called *round dance* and the *waggle dance*. During the round dance, the bee runs around in a little circle first one way, then turns around and runs in the opposite direction of the circle (Fig. 2.3A). The round dance communicates that the food source is near the hive (within 10 - 80 yards) and of its taste and smell, but gives no information about the direction of the food source (Hamdan, 1997; Howland, 2009).

For a food source found at a greater distance from the hive, the worker bee performs the waggle dance by shivering side-to-side motion of her abdomen on the surface of the comb. While dancing, she forms a figure eight (Fig. 2.3B) by moving along a straight line on the comb to some distance and turning to one side to return to the starting point. This sequence is then often repeated over 100 times, with the direction of the return phase circling alternating each time. The vigor of the waggle, the number of times it is repeated, the direction of the dance, and the sound the bee makes communicates amazingly precise information about the location and abundance of the food source. The duration of the waggle phase is correlated to the distance of the food source and the number of cycles performed is correlated to the size of the food supply. The further the foraging site, therefore, the longer the duration of the waggle, and the bigger the food source the greater the number of dance cycles (Howland, 2009). During this time, the bee transmits the waggle vibrations to the surface of the comb, where they are felt by other bees attending the dancer, especially if the dance is performed on uncapped cells (Diana and Alphonse, 1998).

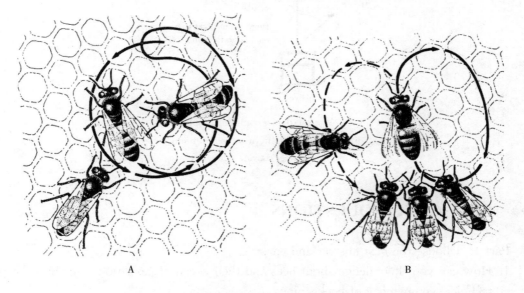

Fig. 2.3 Bees' Round (A) and Waggle (B) dance communication on comb

The angle of the straight line from the vertical (vertical comb) is equal to the angle between the food source and the sun upon departure from the hive, and the vigour with which the waggle is performed is an indication of how much food is present at the site (David, 2008). If the nectar source and the sun are in the same direction, the bee runs the straight line between the two half-circles with her head up towards the sun indicating to the bees 'you can find the nectar by flying toward the sun'. If the nectar source and the sun are in opposite direction the bee runs the straight line with her head down indicating the nectar source is opposite the direction of the sun. If the nectar is at the right or left of the sun, the bee orients the straight run of here dances at the right or left of the vertical on the comb (Hamdan, 1997) (Fig. 2.4).

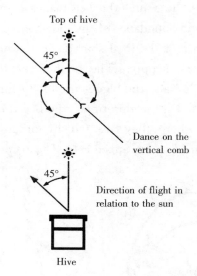

Fig. 2.4　Bees' Waggle-dance communication on comb showing
direction of their flight in relation sun
Source: David, 2008.

>>> SELF-CHECK QUESTIONS

Part 1. Choose the best answer and encircle.
1. How can your knowledge about bees and their social organization contribute for your technical competency of beekeeping?
   A. Enables to handle the bees and work with them safely
   B. Does not has any significant contribution
   C. It reduces one's fear to approach and work with bees.
   D. Enables to identify and use beekeeping tools and equipment
2. How can queen bee physically distinguished from other members of the colony?
   A. By her much shorter body than that of worker's and drone's
   B. Her wings fully cover her abdomen.
   C. Viewed from the front, her head is triangular
   D. By her wasp like slender abdomen
3. The queen bee is the heart and soul of the colony with the responsibility of:
   A. Egg laying
   B. Producing a pheromone that unifies and guide the colony
   C. Caring for her young
   D. A & B
4. One of the followings indicates the presence of queen in the colony.
   A. Presence pollen grain in the comb　　B. Freshly laid eggs

MODULE 2
RECOGNIZE HONEY BEE COLONY AND ITS ORGANIZATION

    C. Presence of pupae   D. Abundance of drone in the hive

5. Queen bee uses her pheromone to
   A. Induce worker bees to lay eggs
   B. Inhibit the development of worker's ovary
   C. Enhance the larval development of brood
   D. Determine the drone bee with which to mate

6. The fertilized eggs laid by queen bee develop into:
   A. Worker bee and drone   B. Drone
   C. Queen and drone   D. Either worker or queen

7. The duties of worker bees for the first 21 days of her life include:
   A. Fetching water   B. Defending the hive
   C. Hive cleaning and caring for larvae   D. Collecting pollen and nectar

8. Which of the following is true about drone bee?
   A. Its main responsibility is mating the queen.
   B. Its abdomen is longer than that of queen's.
   C. Sometimes help workers in defending the hive.
   D. Its development period is 21 days.

9. By their round dance, the bees communicate that:
   A. The food source is near the hive
   B. Their enemy is near by their hive
   C. The food source is found at longer distance from their hive
   D. Absence of forage

Part 2. Write the correct answers of the following questions.

1. How do bees from different hives know each other?
2. Why worker bees can't lay eggs in the presence of queen?
3. What freshly laid eggs look like? And how about larva?
4. How can you know the presence of queen bee without locating her in the colony?
5. If a forager bee discovers the forage at far distance from her hive, how can she tell other bees?

>>> REFERENCES

Blackiston H, 2009. Beekeeping for Dummies [M]. 2nd ed. Wiley Publishing, Inc., Indianapolis, Indiana.

Sammataro D, Avitable A, 1998. The Beekeeper's Hand Book [M]. 3rd ed. Cornell University Press, USA.

Segeren P, 2004. Beekeeping in the tropics [M]. 5th ed. Digigrafi, Wageningen, the Netherlands.

MODULE 3: IDENTIFY THE STRUCTURE, LOCATION AND FUNCTION OF HONEYBEE'S BODY PARTS

>>> INTRODUCTION

Honeybees are well equipped for life as social animals to be successful in the environment. As in most other insects, the honeybee body consists of three anatomical sections or regions: the head, the thorax and the abdomen. In addition, honey bees body parts are covered with exoskeleton and body hair. Internally, honey bees have different organs and systems that are helpful for proper function in their lives. Thus, understanding their anatomy/body parts and their functions can help beekeepers understand bee biology. This module enables to identify and describe external and internal body parts and their functions in honeybees.

1 COMPONENTS AND FUNCTIONS OF EXTERNAL BODY PARTS

As in most other insects, the honeybee body consists of three anatomical sections or regions (Fig. 3.1): 1) *the head region*: with mouthparts and sensory organs such as eyes and antennae; 2) *the thorax*: a locomotor center which is almost entirely filled with muscles that operate the membranous wings and jointed legs, and 3) *the abdomen*: more spacious than the other parts, which holds the organs for various functions, including digestion, circulation, reproduction, and stinging. Honey bees body parts are covered with exoskeleton and body hair (Dade, 1994; Seeley, 1995; MAAREC, 2004; Huang, 2009; Warren, 2014).

1.1 Head Region

The honey bee head is triangular when seen from the front. The head region is the forward body region of the honey bees that contains: antennae, eyes (compound and simple eyes), and mouth parts (mandibles and proboscis) as shown in Fig. 3.2. The head is the centre of information gathering. It is here that the visual (sight), gustatory (taste) and olfactory (smell) inputs are received and processed. The head region is also the site for ingesting food and partially digests food through the mouthparts and associated glands, and serves as the major sensory region of the body through the eyes, antennae, and sensory hairs (Huang, 2009).

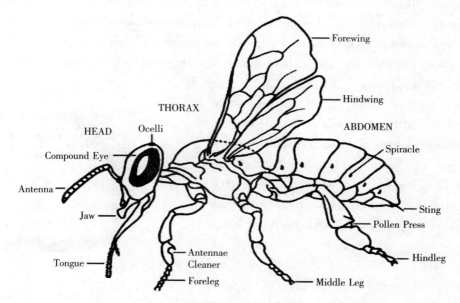

Fig. 3.1 External body parts and regions of honeybee
Source: Huang, 2009.

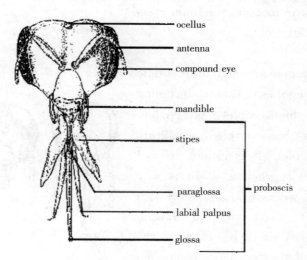

Fig. 3.2 The head region parts of honeybee
Source: J. Ellis, J. Graham, C. Nalen, 2014.

1.1.1 Antennae

In honey bees, the segmented antennae are important sensory organs. The antennae can move freely since their bases are set in small socket-like areas on the head and are bent. This almost 90° 'elbow' allows for the honeybee to have a greater radius for 'feeling' its surroundings. It can turn the antennae at that joint and have contact with more area than a straight antenna would offer. The antennae of the honeybee are extremely sensitive and have thousands of sensory organs, some are specialized for touch (mechanoreceptors), some for smell (odour receptors), and others for taste or gustatory receptors (Dade, 1994; Goodman, 2003; Huang, 2009).

The antennae have three parts: scape (attaching to the head), pedicel (small joint), and flagellum (where most of the sensory parts are located) as shown in Fig. 3.3. Each of the antennae is connected to the brain by a large double nerve that is necessary to accommodate all the crucial sensory input. The tiny sensory hairs on each antenna are responsive to stimuli of touch and odor.

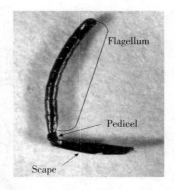

Fig. 3.3 Honeybee antennae
Source: Huang, 2009.

The main function of the antenna of a honey bee are:
- Sense of smell: The antenna can detect odors and even the direction of the odor.
- Sense of touch (feel)
- An instrument to measure the flight speed
- Detect flight direction

1.1.2 Eyes

Honey bees can perceive and differentiate between six major categories of colors, including yellow, blue-green, blue, violet, ultraviolet, and a color known as 'bee's purple', a mixture of yellow and ultraviolet. Bees cannot see red color, it makes them aggressive. That is why they attack when wear red colored cloth (MAAREC, 2004). According to Huang (2009), honeybees have two types of eyes: the three simple eyes and the two large compound eyes. There is a difference in the relative size and position of the two types of eyes in different casts as shown in Fig. 3.4.

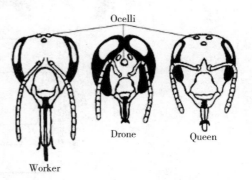

Fig. 3.4 Honeybee eyes
Source: Z. Huang, 2009.

MODULE 3
IDENTIFY THE STRUCTURE, LOCATION AND FUNCTION OF HONEYBEE'S BODY PARTS

Simple eyes

There are three simple eyes, called *ocelli*, located at the top of the head between the bee's two larger compound eyes. These eyes work to triangulate the bee's position related to the sun and act as a sort of navigation system for it. This allows a honeybee to know the way home once it has completed a few initial orientation flights when first leaving its hive. This is because the ocelli detect light but can't focus or arrange an image like the larger compound eyes. Ocelli register intensity, wavelength, and duration of light. At dusk the ocelli estimate extent of approaching darkness, causing the bees to return to their hives.

Compound eye

Honey bees have two compound eyes that make a large part of the head surface. Each compound eye is composed of individual cells (facets) also called ommatidium, plural ommatidia. Each facet (ommatidium) is composed of many cells, usually including light focusing elements (lens and cones), and light sensing cells (retinal cells). Facets are basic unit of sight. Workers have 4,000 – 6,000 ommatidia but drones have 7,000 – 8,600, presumably because drones need better visual ability during mating.

> Compound eyes respond independently to incoming light waves and groups of facets are specialized for receiving 1) polarized light, 2) pattern recognition, 3) color vision, and 4) head turning responses (orientation).
> Compound eye can also perceive airflow using sensory hairs at the junction of facets.
> Removing sensory hairs disrupts bees' ability to estimate flight distance.
> Facets diverge angularly by 1° resulting in a mosaic pattern that is well adapted to detecting movement.

1.1.3 Mouth

Honey bees have a combined mouth parts than can both chew and suck. The mouth part comprises of mandible and proboscis (Huang, 2009) as shown in Fig. 3.2.

Mandibles

The mandibles are the paired, spoon-shaped, strong and jaw-like 'teeth' that can be open and closed sideways. Powerful muscles connect them to the head. Functions of mandibles are:

> *Eating*: Honeybees use their mandibles to eat 'bee bread', a mix of fermented pollen and nectar.
> *Fighting*: Honeybees bite invaders, mites, robbers and other threatening creatures that approach or try to enter their hive.
> Shaping and trimming (manipulating) beeswax in the hive into hexagonal honeycomb
> Shaping and manipulating propolis for nest construction

- Feed brood food to larvae and nectar to the queen
- Drag debris and the dead from the nest
- Grooming each other

Proboscis (tongue)

The proboscis is in front of the mandibles. The proboscis of the honey bee is simply a long, slender, hairy tongue that acts as a straw to bring the liquid food (nectar, honey and water) to the mouth. When in use, the tongue moves rapidly back and forth while the flexible tip performs a lapping motion. After feeding, the proboscis is drawn up and folded behind the head.

The proboscis has four parts: stipes, paraglossa, labia palpus and glossa. There is a difference in size and length of proboscis among the casts, it is larger and longer and well developed in workers, intermediate in case of queen, but shortest in case of drones.

Proboscis mainly used for sucking in liquids such as nectar, water and honey inside the hive, for exchanging food with other bees (trophallaxis), licking to transfer pheromones or grooming and also for removing water from nectar. Movement of liquids occurs by back and forth movements of the glossa, capillary action, and pumping of the muscles of the cibarium (hollow area at the base of the glossa) which creates suction for ingestion. Proboscis indirectly collects pollen while foraging and then groomed from the proboscis to the forelegs.

1.2 Thorax Region

The thorax region is the middle section of the honey bees' three regions that contains the two pair of wings and three pairs of legs. The thorax is the center for locomotion.

It is divided into three segments: prothoracic, mesothoracic and metathoracic segments (Ellis et al., 2014; Warren, 2014).

- *Prothorax* is the first thoracic segment and is where the forelegs are attached. There is no wing attachment on the prothorax.
- *Mesothorax* is the second and middle thoracic segment. This is the point of attachment for the forewings and the middle pair of legs.
- *Metathorax* is the third and final thoracic segment. This is the point of attachment for the hind legs and hind wings.

1.2.1 Wings

Honeybees have two pairs of wings (the front/fore wings and the hind wings). The front wings are larger than the hind wings and the two are synchronized in flight with a row of wing hooks (humuli, singular: humulus) on the hind wing that would hitch into a fold on the rear edge of the front wing. Each vein in the wing has a name and wing venation is a key character used to differentiate European from African honey bees.

The wings are powered by two sets of muscles inside the thorax, the longitudinal and vertical muscles. During flight, when the longitudinal muscles contract, the thorax raises

MODULE 3
IDENTIFY THE STRUCTURE, LOCATION AND FUNCTION OF HONEYBEE'S BODY PARTS

its height, so the wings are lowered because of fulcrum like structure (pleural plates) near the wing base. Conversely, when the vertical muscles contract, it shortens the height of thorax, raising the wings. The honey bee flight muscles can contract several times with one single nerve impulse, allowing it to contract at a faster rate.

1.2.2 Legs

Honeybees have three pairs of legs. One pair per thoracic segment and their basic construction is the same as shown in Fig. 3.1. The legs are very versatile, with claws on the last tarsomere, allowing bees to have good grip on rough surfaces like tree trunks, but also with a soft pad (arolium) to allow bees to walk on smooth surfaces (leaves or even glass). There are also special structures on legs to help bee get more pollen. Parts of the typical leg (Fig. 3.5 A) are: femur (segment closest to the thorax), tibia (attached to the femur), basitarsus (attached to the tibia), and tibia or foot part (claws and arolium).

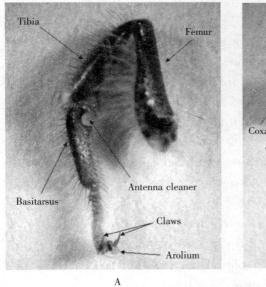
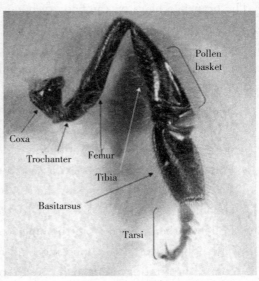

Fig. 3.5 Front and hind leg of honey bees
Source: Huang, 2009.

Front legs are also called antenna cleaner - on the front leg, there is a special structure used for cleaning antenna (when too many pollen grains stuck there), properly called the 'antenna cleaner'. A curved notch and associated spur through which the antenna is pulled and brushed clean. The front legs are hairy and used in general as a brush for cleaning (Fig. 3.5 A).

Middle legs have no special modifications like the front and hind legs. They are hairy and are used for cleaning.

Hind legs are highly modified for pollen and propolis collection in worker bees (Fig. 3.5 B). Prominent structures on hind legs for pollen and propolis collection are: pollen

basket or corbicula, pollen press and comb.
- *Pollen basket* (s) also called a corbicula; it is a smooth, somewhat concave surface of the outer hind leg that is fringed with long, curved hairs that hold the pollen in place. This enclosed space is used to transport pollen and propolis to the hive.
- *Pollen press*: Once the bees have gathered the pollen, they move it to the pollen press located between the two largest segments of the hind leg. It is used to press the pollen into pellets.
- *Rakes and combs* are structures on the legs used to collect and remove pollen that sticks to the hairy bodies of honey bees.

Pollen collection processes in worker bees

The branched hairs on the body of honey bee are adapted to collect pollen. Worker bees gather/collect pollen by two methods: 1) by active movements of the legs and proboscis scraping the anthers, 2) passively shed pollen onto body hairs (Huang, 2009). The pollen collection process is as follows:
- The forelegs brush the proboscis, picking up pollen made sticky by regurgitated honey, and gather pollen from the head and front of thorax.
- When airborne the worker hovers, transferring pollen from the forelegs and the posterior thoracic segments to the middle legs.
- Pollen on the middle legs is passed to the pollen combs on the inner hind legs by scraping the middle legs past the combs; the combs also pass the pollen from the abdomen.
- Pollen is then passed from the inner combs to the outer corbicula.
- The pollen rake of the opposing hind leg scrapes the inner surface of each pollen comb, facilitating the transfer to the outer surface of the leg.
- Bees force the pollen into an area called the pollen press.
- Pollen is forced into the corbicula by pumping motions to the hind legs against each other.
- After packing in the pollen basket, the worker makes flight to their hives. Pollen foragers deposit their loads in to cells without assistance.

Propolis collection processes

- Propolis is a resinous gum, collected by bees with their mandibles from trees or other plants. The corbiculae are used to transport propolis, but the pollen presses do not play a role in loading in to the corbiculae.
- Resin is transferred from the mandibles to the fore and middle legs and placed directly into the corbiculae, for transportation.

MODULE 3
IDENTIFY THE STRUCTURE, LOCATION AND FUNCTION OF HONEYBEE'S BODY PARTS

1.3 Abdomen Region

The abdomen is the center for most of the vital organs in the honey bee. The honey bee abdomen is composed of ten segments. The first seven are visible segments and the last three segments are modified in to reproductive organ and sting. It is the center for digestion and reproduction (for drones and queens). It also houses the sting, a powerful defence against enemies. A worker, a drone, and a queen can be distinguished by the size and shape of their abdomen. The queen pictured here was laying eggs, as can be seen by her large abdomen. External body parts on the abdomen are: wax scale (glands), spiracles and sting (Winston, 1991; Dade, 1994; Huang, 2009; Ellis et al., 2014).

1.3.1 Wax Scales

Worker bees of 6 - 12 days old can produce wax scales on the ventral abdominal segments. There are four pairs of wax glands to secrete wax. The glands are concealed between the inter-segmental membranes, but the wax scales produced can be seen, usually even with naked eyes.

The wax is discharged as a thin and clear liquid and hardens to small flakes or scales and sits in wax pockets. The worker bee draws the wax scales out with the comb on the inside hind leg. The wax scale is then transferred to the mandibles where it is chewed, and saliva is added into a compact, pliant whitish mass. The beeswax is then added to the comb. After the worker bee outgrows the wax forming period, the glands degenerate and become a flat layer of cells.

1.3.2 Spiracles

Spiracles are external respiratory openings of honeybees located on the ventral side of the thorax and abdomen. They are breathing holes, a site for gas exchange (entry of oxygen and exit of carbon dioxide) in honeybees. There are 3 pairs on the thorax and 7 pairs on the abdomen.

1.3.3 Sting

The honey bee sting is a modified ovipositor used for defense instead of for laying eggs (Fig. 3.6). The worker honey bee's sting is barbed and stays beneath the skin after the doomed bee flies away. The purpose of a sting is for defense, however once a honey bee stings, it also loses its life. The worker leaves the sting in the body of the victim and when pulling away ruptures the abdomen and thus dies (Fig. 3.6A). However, the sting of queen is not barbed at the tip and used to kill rival queen, and can be retracted after stinging, and not lethal for the queen (Fig. 3.6B).

The sting is found in a chamber at the end of the abdomen, from which only the sharp-pointed shaft protrudes. It is about 0.32cm long. When the stinger is not in use, it is retracted within the sting chamber of the abdomen. The shaft is turned up so that is base is concealed. The shaft is a hollow tube, like a hypodermic needle. The tip is barbed so that it sticks in the skin of the victim (Fig. 3.6A). The hollow needle actually has three sections. The top section is called the stylet and has ridges. The bottom two pieces are called lancets. When the stinger penetrates the skin, the two lancets move back and forth on the ridges of the stylet so that the whole apparatus is driven deeper into the skin. The poison canal is enclosed within the lancets.

In front of the shaft is the bulb. The ends of the lancets within the bulb are enlarged and as they move they force the venom into the poison canal, like miniature plungers. The venom comes from two acid glands that secrete into the poison sac. During stinging, the contents of the alkaline gland are dumped directly into the poison canal where they mix with the acidic portion.

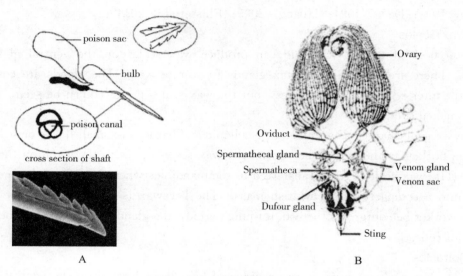

Fig. 3.6 Sting of worker and queen bee
A. Sting of worker bees with shaft B. Stinger of queen bee

When a honey bee stings a mammal, the stinger becomes embedded. In its struggle to free itself, a portion of the stinger is left behind. This damages the honey bee enough to kill her. The stinger continues to contract by reflex action, continuously pumping venom into the wound for several seconds.

If the stinger squeezed more and more venom is released in to the victim wound that increases its effect. For this reason, one should not 'pull out' a honeybee stinger. They will inadvertently squeeze out more venom directly into the wound. Using a stiff-edged object to lift the stinger out from the bottom of the barbs is less likely to empty the venom sac.

Alarm pheromone is released to 'mark' the victim from the sting gland that sends a chemical signal through the air to other honeybees in the vicinity that danger is near. This puts those honeybees on guard and the rest of the hive is warned (MAAREC, 2004).

1.4 Exoskeleton of Honey Bees

Honey bees, like all insects, have segmented exoskeleton which is rigid and covered with layers of wax, but have no internal bones like vertebrates do. Honey bees are segmented in nearly all their body parts: three segments of thorax, six visible segments of abdomen (the other three are modified into the sting or reproductive organ), legs and antennae are also segmented. The main component of exoskeleton is chitin which is a polymer of glucose and can support a lot of weight with very little material (Huang, 2009). Functions of the exoskeleton are:

> The exoskeleton protects their vital organs from the environment and provides areas of attachment for muscles and connective tissue.
> The wax layers protect bees from desiccation (losing water).
> The exoskeleton also prevents bees from growing continually; instead, they must shed their skins periodically during larval stages, and stay the same size during the adult stage.
> Exoskeleton allows rapid but precise movements (moving in a suit of armour).
> Protection from predators

1.5 Body Hair

The honey bee body part is covered in small branched hairs called plumose, even between the facets of the compound eyes. Bees are distinguished from other Hymenoptera by these plumose hairs. These hairs have sensory capabilities, enhance pollen collection/transportation ability and help protect the exoskeleton keeping it free of debris (Ellis et al., 2014).

2 COMPONENTS AND FUNCTIONS OF INTERNAL BODY PARTS

The internal anatomy of the honey bee is a network of organ systems and muscles. Those includes: digestive system, nervous system, circulatory system, reproductive system, respiratory system and endocrine system (Fig. 9).

2.1 Digestive Organs and System

The digestive tract of honeybees is rather typical for an insect. The digestive system also called the alimentary canal is responsible for food intake, nutrient uptake and excretion. The alimentary canal includes: mouth, oesophagus, salivary glands, honey crop (honey stomach), proventriculus, ventriculus (true stomach/midgut), Malpighian tubules, small intestine (ileum), and rectum. Honey bees have reversible movement of foods from mouthparts to a honey stomach or crop (Winston, 1991).

Mouth: Food uptake is by mouth and some digestion takes place here by glandular sections such as salivary glands. Food passes from mouth to esophagus.

Oesophagus: The oesophagus starts near the mouth, goes through an opening in the brain, through the thorax, and enlarges near the end to form the honey crop. From the oesophagus, food enters the crop or honey stomach.

Crop (honey stomach): The honey stomach is a crop or storage area to hold freshly collected nectar (or water) for transport and then deposit inside the nest following regurgitation.

Proventriculus is a special structure near the end of the crop. It has sclerotized teeth-like structure, and muscles and valves. These structures allow the workers to remove pollen grains in the nectar, and stopping the backflow of food being digested into the crop, ensuring that the nectar is never contaminated. The contents of the crop can be spit back into cells, or feed to other workers, as is the case of nectar collected by foragers.

Ventriculus (midgut) is also called true stomach of honeybees. The ventriculus is the functional stomach of bees and is the largest part of the intestine. Most digestion and absorption occurs in the midgut.

Malpighian tubules are small strands of tubes attached near the end of ventriculus and functions as the kidney, it removes the nitrogen waste (in the form of uric acid, not as urea as in humans) from the hemolymph and the uric acid forms crystals and is mixed with other solid wastes.

Ileum and *rectum*: The alimentary canal is completed by a short small intestine (ileum) and a large intestine (rectum). They compose the hindgut where food digestion is completed. Undigested food residues are reformed into faeces in the rectum and eliminated through the terminal anus. The rectum is quite expandable structures, enabling workers to refrain of defecation for up two months.

2.2 Respiratory Organs and System

Honey bees have external respiratory openings called spiracles. There are 3 pairs on the thorax and 7 pairs on the abdomen. They access the trachea. The trachea arms widen to form air sacs. The small branches and tubes (trachea and tracheoles) emerging from the sacs extend to the tissues. Bees can accelerate the passage of air into their bodies by contracting these sacs, thus speeding the oxygenation of the tissues. Thus, the tracheal system carries oxygen to and carbon dioxide (CO_2) away from cells.

At rest, respiration occurs passively by diffusion. Under stress, such as during flight, bees pump their abdomens to increase gas exchange and expand air sacs of the trachea like bellows, facilitating greater gas exchange (Huang, 2009).

2.3 Circulatory System

The honey bee has an open circulatory system; the main pumping organ of this system is the dorsal blood vessel (dorsal heart and aorta). They assist in blood circulation. The dorsal aorta runs from the abdomen to the head. The dorsal heart is in the abdomen. The blood (hemolymph) of the honey bee is not responsible for oxygen delivery. Hemolymph does not carry haemoglobin; thus, it is not red in colour. Its primary function is the distribution of digested food material and transfer of carbon monoxide, defensive proteins, and waste materials from cells to excretory organs (Dade, 1994).

2.4 Reproductive System

The mature reproductive cells of the male are called spermatozoa, and the female eggs or ova. Ovaries of the queen are paired, each with a bundle of more than 150 ovarioles (Fig. 10). Workers also have paired ovaries, but the number of ovarioles depends on colony conditions. Queens have a spermatheca while worker bees do not.

Ovarioles of a fertile queen are large and fill almost the entire abdominal cavity. When an egg is ready to be discharged, the follicle opens and the egg passes down the oviduct. Individual ovarioles can be seen, with more mature eggs shown as yellowish. The eggs can be fertilized by spermatozoa stored in the spermatheca. Only eggs intended to develop into female larvae are fertilized. Eggs that will become males are not. The spermatheca is shiny, perfectly spherical organ when the tracheal tissues are removed.

In case of drones, testes are composed of tubules in which sperm are produced and mature. At sexual maturity (12 - 13 days post emergence), the testes are reduced. The endophallus is everted on mating while a pair of copulatory claspers grips the queen during copulation. While mating, the endophallus breaks off, is left in the queen, and the drone dies (Ellis et al., 2014).

2.5 Nervous System

The main internal organs in the head are the brain and sub-esophageal ganglion, the main component of the nervous system, in addition to the ventral nerve cord that runs all the way through the thorax to the abdomen. Yes, the bee does have a brain, a pretty sophisticated one too. The brain has a large area for receiving inputs from the two compound eyes, called optic lobes. The next largest input is from the antenna (antenna lobes). One important region in the middle of the brain is called the 'mushroom body' because the cross section resembles two mushrooms. This area is known to be involved in olfactory learning and short-term memory formation, and recently shown to be also important in long term memory formation in insects (Dade, 1994; Huang, 2009).

Generally, nervous system of honeybees:

> Consists of the brain and seven *ganglia* at various junctions throughout the body
> Most locomotion is controlled by the ganglia not the brain.
> A beheaded insect can move its legs and wings vigorously. A decapitated bee can walk and sting, but flying is not possible because it is out of balance without the head.
> The bee brain consists of a small bundle of cells with all the automatic functions transferred to the ganglia. The ganglia are reduced to barely visible proportions.

2.6 Endocrine/Gland System

Four basic functions of glands in honeybees are: wax production, communication [internal (hormones) and external (pheromones) communication], defence and food processing (Caron, 1999).

There are exocrine glands inside the head. These are: the mandibular glands, the hypo-pharyngeal glands and the salivary glands. Other glandular systems in the thorax and abdomen are: wax glands, Nasonov scent glands, sting gland, and Arnhart gland.

Mandibular gland is a simple sac-like structure attached to each of the mandibles.
- Queen: In the queen, this is the source of the powerful queen pheromone. It produces the queen substance. It suppresses the reproductive organs of worker bees and the typical queen pheromone that maintain the integrity of the colony.
- Nurse bees: In young worker bees, the gland produces a lipid-rich white substance that is mixed with the secretion of hypo-pharyngeal glands to make royal jelly or worker jelly and fed to the queen or other workers.
- Foragers: In old workers (foragers), the gland also produces heptanone, a component of the alarm pheromone.

Hypopharyngeal glands: The glands consisted of a central duct (which is coiled between the front cuticle and the brain) with thousands of tiny grape-like spheres. The secretion flows to the mouth through the long duct. The glands are large (hypertrophied) in nurse bees but become generated in foragers.
- In young (nurse) bees, it produces protein-rich secretions called the royal jelly. It is the largest glands in the worker bee and well developed in nurse bees.
- In foragers, it produces invertase (an enzyme to break down sucrose into fructose and glucose).

Salivary glands: There is a pair of salivary glands inside the head. They aid in the digestion of food by secreting an enzyme invertase. The glands produce saliva which is mixed with wax scales to change the physical property of wax.

Wax glands produce wax in segments of abdomen on the fourth through seventh ventral abdominal segments, secrete wax in liquid form onto wax plates. The wax then hardens into wax scales that are used by workers to construct comb.

Nasonov scent gland produces a pheromone that attracts other bees. It orients swarm/hive entrance/flower attraction. It is produced in the last abdomen segment. It is dispersed by wing buzzing.

Sting gland releases alarm pheromones which are used to alert other bees and mark an enemy.

Arnhart gland leaves the 'scented footprint' on flowers and hive. It is produced in the tarsal segment of legs.

MODULE 3

IDENTIFY THE STRUCTURE, LOCATION AND FUNCTION OF HONEYBEE'S BODY PARTS

>>> **SELF-CHECK QUESTIONS**

Part 1. Choose the best answer among the alternatives given.
1. Which of the following is *wrong* about exoskeleton of honeybees?
   A. It is made up of rigid structure called chitin.
   B. It protects bees from desiccation (body water loss).
   C. It provides a site for organ attachment.
   D. It enables for continuous growth of their body in their life.
2. Body hair in honeybees is used as
   A. Sensory organ
   B. Pollen collection and transportation
   C. Protect exoskeleton from debris
   D. All
3. Which of the following is *not* part of the honeybee's head region?
   A. Proboscis   B. Mandible   C. Honey stomach   D. Ocelli
4. Head region of honeybee is
   A. Centre of information gathering
   B. Centre of locomotion
   C. Centre of digestion and reproduction
   D. All
5. Function of antennae in honeybee is
   A. Sense of smell or odour
   B. Detect direction of flight
   C. Sense of taste
   D. All
6. An organ of honeybees used to detect light intensity and duration, causing the bees to return to their hive at dusk
   A. Compound eyes
   B. Simple eyes
   C. Antennae
   D. Plumose hairs
7. Which of the following is *not* function of compound eyes in honeybees?
   A. Perceive air flow
   B. Pattern recognition
   C. Arrange and focus image, and detect object movement
   D. Detect light intensity and duration
8. _____ are mouth parts of honeybees used for chewing and sucking, respectively.
   A. Proboscis and mandible
   B. Mandible and proboscis
   C. Glossa and ocelli
   D. Proboscis and ommatidium
9. Locomotion in honeybees is controlled by
   A. Ganglia   B. Spiracle   C. Brain   D. All
10. Which of the following is *not* function of mandible?
    A. Sucking of nectar/liquid
    B. Chewing food
    C. Manipulation of wax and propolis
    D. Feed food to larva and queen

11. Which of the following is *not* function of proboscis?
    A. Sucking of liquid/nectar/water B. Licking to transform pheromones
    C. Grooming other bees D. Chewing food/bee bread
12. Thorax of honeybees has
    A. Three segments
    B. Three pairs of spiracles
    C. Three pair of legs and two pair of wings
    D. All
13. Pollen basket in worker bees is located at
    A. Fore/front legs B. Middle legs
    C. Hind legs D. All
14. Which of the following is *not* function of circulatory system in honeybees?
    A. Transport oxygen and carbon dioxide
    B. Transport food
    C. Transport hormones
    D. All
15. Glandular system in honeybees is/are used for
    A. Wax production B. Communication and defence
    C. Food processing D. All
16. Which of the following casts have large and longer proboscis?
    A. Drones B. Queen C. Worker D. None
17. Which pair of legs of honeybees is also called antennae cleaners?
    A. Fore/front legs B. Middle legs C. Hind legs D. None
18. Wax gland in worker bees is in the _____ region/segment of the bees.
    A. Head B. Thorax C. Abdomen D. All
19. Which of the following is correct about abdomen of honeybees?
    A. Abdomen have seven segments.
    B. Queen bees have larger and longer abdomen.
    C. Drone abdomen is loner and broader than worker bees.
    D. All
20. Which of the following casts do not have sting gland?
    A. Drone B. Worker C. Queen D. None
21. Function of Nasonov gland is
    A. Orientation of swarm, hive entrance and flower attraction
    B. Make alarm
    C. Leaves the scented foot print on hive and flowers
    D. Produce invertase for food digestion
22. Hypo-pharyngeal gland produce
    A. Alarm pheromone B. Royal jelly

IDENTIFY THE STRUCTURE, LOCATION AND FUNCTION OF HONEYBEE'S BODY PARTS

   C. Invertase   D. All
23. The function of Arnhart gland is
    A. Orientation of swarm, hive entrance and flower attraction
    B. Make alarm
    C. Leaves the scented foot print on hive and flowers
    D. Produce invertase for food digestion
24. The function of the secretion of sting gland is
    A. Orientation of swarm, hive entrance and flower attraction
    B. Leaves the scented foot print on hive and flowers
    C. Make alarm
    D. Produce invertase for food digestion
25. The salivary gland
    A. Orientation of swarm, hive entrance and flower attraction
    B. Make alarm
    C. Leaves the scented foot print on hive and flowers
    D. Produce invertase for food digestion

Part 2. Match column A with the appropriate words/phrases from column B.

A		B
1. Humuli	A	Blood of bees
2. Prothorax	B	Function as kidney and liver in bees
3. Metathorax	C	Site for breathing/gas exchange
4. Mesothorax	D	True stomach of honeybees
5. Ocelli	E	Enable bees to walk on smooth surface
6. Ommatidium	F	Sperm storage
7. Spiracle	G	Facet/basic unit of eye
8. Corbicula	H	Wing hooks
9. Hemolymph	I	Comprise middle legs and fore wings
10. Malpighian tubule	J	Comprise hind legs and hind wings
11. Spermatheca	K	Comprise fore/front legs
12. Ventriculus	L	Food exchange among bees
13. Crop	M	Simple eyes
14. Trophallaxis	N	Pollen basket
15. Arolium	O	Honey stomach

Part 3. Say *'True'* if the statement is correct or *'False'* if the statement is incorrect.

1. The hind wings are larger than the fore/front wings.
2. Queen and worker bees have stings.
3. Worker bee sting can be retracted after stinging, but queen bee sting is barbed and stay breath skin after sting.
4. Queen bee can produce royal jelly.
5. Honey bees have closed circulatory system.

Part 4. Discussion questions.

1. Draw a sketch of honeybee and label the external body parts.
2. Discuss the pollen collection procedures by worker bees.
3. Briefly describe digestive tracts and process in honeybees.
4. How do honeybees breathe?
5. Compare anatomical differences among honeybee casts (worker, drone and queen).

>>> REFERENCES

Caron D M, 1999. Honey bee biology and beekeeping [M]. Wicwas Press, Cheshire, CT.
Dade H A, 1994. Anatomy and Dissection of the Honeybee [M]. Alden Press, Oxford, UK.
Seeley T D, 1995. the wisdom of the hive: the social physiology of honeybee colonies [M]. Harvard University Press.
Winston M L, 1991. Biology of the Honey Bee [M]. Harvard University Press.

MODULE 4:
IDENTIFY AND PREPARE BEEKEEPING EQUIPMENTS AND TOOLS

>>> INTRODUCTION

If you want the start of your beekeeping career to go smoothly, make sure whether you have all necessary equipment and tools such as suitable bee hives, clothing, smokers, hive tools and other tools. These equipment and tools can be made from locally available resources or can be obtained by purchasing from local suppliers. In general, beekeeping equipment can be classified under different heading such as: 1) hives and their accessories, 2) protective clothing to prevent sting, 3) tools used in opening, harvesting, manipulating and transporting hives, 4) equipment used in handling the crops of honey and beeswax. This module enables beekeepers to identify and prepare beekeeping equipment and tools as their categories.

1 BEE HIVES AND THEIR ACCESSORIES

There are three types of bee hives in Ethiopia, namely traditional backyard, transitional and improved beekeeping or modern hive (Holeta Bee Research Center, 2004).

1.1 Traditional Hive (Fixed Comb Hives)

Fixed comb hives (traditional hives) are containers made from whatever materials those are locally available, such as grasses, logs, bark, raffia palm, clay. Bees build their nest inside the container, just as they would build in a naturally occurring cavity. The bees attach the combs to the inside upper surface of the hive. The honeycombs need to be cut off from this surface to be harvested and cannot then be replaced (FiBL, 2011). Fixed comb hives such as the hollowed-out logs, bark hives, clay pots and woven grasses, etc. are cheap to construct, relatively easy to manage and suitable for defensive bees in tropical Africa, including Ethiopia (Tessega, 2009).

Advantages of traditional hives

➢ Cheap
➢ Materials are locally available
➢ Does not require a lot of skills and technology
➢ High propolis productivity
➢ High wax productivity

Disadvantages of traditional hives

➢ Difficult to inspect for harvesting and brood manipulation
➢ Combs break when transported over long distances
➢ Production is limited since hive cannot be extended (has no possibilities of supering)
➢ Loss brood while harvesting
➢ Difficult to harvest and a lot of smoke is needed
➢ Difficult to determine harvesting capacity or volume because of differences in length and diameter
➢ Swarming and absconding are common

Types of traditional hives

The main inputs are local knowledge and local materials, rather than external financial support and donated equipment. Fixed comb hives, usually cylindrical in shape, have been used in Africa and in Ethiopia for generation. A variety of different styles of traditional hive can be found across the continent based on the material they made from, for instance, hollowed-out logs and bark formed into cylinders, clay pots and woven grasses. Local methods have evolved over a long period to suit local resources and indigenous bees. Hive used will depend on the material available in the area. Some type of tradition hive is listed below:

1) Woven basket hive

They vary in shape, size and type of materials used. For example, they can be conical or cylindrical in shape, the cylindrical one measures 30 – 40cm across and 1m long (Yetimwork, 2015). Durability of the hive depends on the materials used and management. One end completely closed, one end bearing 5 – 6 holes of diameter 8 – 10mm in a row. Materials for preparation are: bamboo, fibre, twigs or sticks, cow dung or soil for smearing, grass or banana fibre or dry banana leaves as cover (Fig. 4.1).

2) Log hive

Tree trunk, which is cut into section and hollowed out which is cylindrical in shape and one end closed but one end bears the entrance hole for the bees (Fig. 4.2).

Fig. 4.1 Bamboo woven traditional hive
Source: https://www.wikimedia.org/

Fig. 4.2 Log hive
Sources: MAAIF, 2012.

Advantages

- Cheap
- Materials are locally available.
- Does not require a lot of skills and technology
- High wax productivity
- Durable with good practices
- High colonization rate

Disadvantages

- Difficult to inspect
- Combs break when transported over long distances.
- Production is limited since hive cannot be extended.
- Difficult to harvest and a lot of smoke is needed
- Difficult to determine harvesting capacity or volume because of differences in length and diameter
- Swarming and absconding are common.

3) **Clay or mud jar and bricks** (clay hive)
- Made from baked clay soil
- Can take the cylindrical or oval shapes
- The cylindrical has entrances at one end
- The oval shape has entrances at the bottom with the top covered a plank of wood (Fig. 4.3).

Fig. 4.3 Clay hive
Sources: Ministry of Agriculture, 2012.

1.2 Top-bar (transitional) Hives

Preparation of timber top-bar hives and frames requires advanced machine, which may not be available at farmer's level. Moreover, its cost is relatively high. As a result the dissemination of this hive is very low. To solve these problems in part, preparation of top-bar hives from locally available materials (mud, bamboo, and eucalyptus globules sticks) have been tested and found to be successful without having significant differences with that of machine made hives and frames.

Top-bar hives are removable comb hives made from properly sawn timber with bar placed on the top of hive boxes. They are long boxes carrying a number of planks on top, called top-bars. The bees are expected to build a comb down from each top-bar (FiBL, 2011). Bees are encouraged to construct their combs from the undersides of these top-bars. Top-bars enable the beekeeper to lift individual combs out of the hive for inspection. Combs containing unripe honey or brood can be replaced and those containing ripe honey can be removed for harvest.

Harvesting honey and beeswax from top-bar hives is simple and can be achieved without damage to the colony. Top-bar hives are particularly suitable for beginner beekeepers because it is often easier to learn how to manage and harvest from a top-bar hive than from a fixed comb hive. Installed at waist height and kept close to home, top-bar hives are often popular with women. All the equipment needed for top-bar hive beekeeping can be bought or made locally. Top-bar hives are often introduced by projects keen to promote new and seemingly modern ideas, yet they function well only if the beekeeper understands the bees, the benefits and limitations. The most common types of top bar or transitional hive widely used in tropical Africa including Ethiopia are Kenyan top bar hive, Tanzanian top bar hives, mud block hives and Chefaka hives.

Types of top bar hives

1) Kenyan top bar

It is usually trapezium shaped hive. All top bar hives should have top bars with dimension of width of 3.2cm and a length of approximately 48cm. Kenya top bar hive can be made out of bricks, timber, bamboo, basket & clay.

The cover can be fabricated provided. It is waterproof material, e.g. plastic sheets, grass, banana fibers, and mats (Fig. 4.4).

Fig. 4.4 Kenyan top bar hive
Source: http://peacebeefarm.blogspot.com

Dimensions for body

- Lid: 94cm × 52cm
- Side: 26cm × 87cm
- Bottom: 24cm × (87 - 90) cm
- Ends (2): 25cm in height, 42cm wide at top and 20cm wide at bottom

Dimensions for top bars

- Length: 45.5cm
- Width: 3.2cm

2) Mud block

The body of mud hive can be constructed from mud block. The blocks can be produce from fermented mud using wooden moulds. For construction of mud hive blocks, it requires five moulds to produce different parts. These are bottom and sidewall parts; bottom, front

and rear parts; sidewall supporters; large block and small block.
- ➢ The dimension of the bottom and sidewall mould is 30cm wide × 40cm length with 4cm depth; it requires 6 elements per hive which used to construct 2 each for sidewall and two for bottom.
- ➢ Bottom, front and rear part mould, 60cm length × 27cm width with 4cm depth; elements per hive 1 bottom center, 1 front and 1 rear both with groove
- ➢ Side wall support (triangular) 26cm × 11cm × 23cm with 7cm depth. Two elements per hive
- ➢ Mud block mould 23cm × 19cm with 13cm depth, 16 elements per hive
- ➢ Corner stone mould used to make corner 23cm × 9cm with 13cm depth

3) Chefaka hive (Ethio-ribrab beehives)

This kind of hive is a type of transitional hive made from locally available thin woody plant like the shape of Kenyan top bar and sealed with mud or clay. Ethio-ribrab beehives can be equipped with queen excluder which makes differ from Kenyan top bar hives (Fig. 4.5).

Fig. 4.5 Ethio-ribrab beehive
Source: Gebreamlak et al., 2015.

It requires at least the following local and industrial materials for construction:
- ➢ Eucalyptus of 4 – 5cm diameter or any other suitable wooden poles and nails
- ➢ Queen excluder and plywood
- ➢ Hollow bamboo, plastic cover sheet, wire and rope

Steps of hives construction recommended by IRLI and Ethiopian agricultural research center's bee researchers

1) Cutting wooden materials

Cutting eucalyptus into six different lengths as indicated in Fig. 4.6. The required lengths and quantities of eucalyptus poles include:

MODULE 4
IDENTIFY AND PREPARE BEEKEEPING EQUIPMENTS AND TOOLS

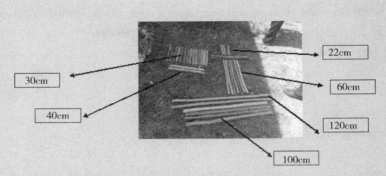

Fig. 4.6   Properly cut framers

- Two pieces of 120cm long: Used as the long side of hive cover
- Four pieces of 100cm long: Used as top and bottom length of hive
- Seven pieces of 60cm long: Used as the short side (width) of hive cover
- Two pieces of 40cm long: Used as top width of hive
- Fourteen pieces of 30cm long: Used as vertical frames/raisers
- Six pieces of 22cm long: Used as bottom width frame

2) Assembling

Fig. 4.7 shows joined eucalyptus poles using appropriate size nails. The gap between the vertical stands is 10, 25, and 30cm.

Fig. 4.7   Attached wooden pole
Source: Gebreamlak et al. 2015.

Once the wooden pole are joined, the dissected hollow bamboo is wave on the five side and bottom poles. The bamboo which is waved on the pole should dissect vertically up to 3cm diameters.

3) Mud plastering and creating a grove for inserting queen excluder/plywood

The inner side of the assembled hive should be plastered using mud composed of loam soil and fine straw, and slowly dried under shade (Fig. 4.7). It is advisable to finish the plastering using cow dung and rub the inner side using plants preferred by bees. This has the advantage of rapid familiarization of bees with the new hive and minimizing absconding.

4) Constructing hive cover and making frames

The hive cover is made from eucalyptus poles, intact bamboo and plastic or iron cover sheets and is 120cm long and 60cm wide. The frame to which the comb is attached should made from up to 22 thick hollow bamboos or other wooden materials (Fig. 4.8).

Fig. 4.8 Wooden frame (top bars)

1.3 Modern Hives (Improved Frame Hives)

Frame hive is standard wooden box which equipped with movable frames.

Advantages of modern hives

- Transportable and long lasting
- High honey yield
- Easy to inspect and harvest
- Easy to control swarming
- Bee breeding and queen rearing possible
- The quality of honey is high due to queen excluder, centrifugal honey extractor and honey strainer used.

MODULE 4
IDENTIFY AND PREPARE BEEKEEPING EQUIPMENTS AND TOOLS

> **Disadvantages of modern hives**
> - Very expensive
> - Some of the materials for construction need to be imported.
> - Requires high skills and technology
> - Production of other hive products is very minimal (wax and propolis).
> - Requires high management skills
> - It is prone to pest and disease attack.

Types of modern hives

Modern hive is the most developed and productive type of hive in the history of beekeeping. In modern type of bee keeping, different types of frame hives is used. Types of modern hives include: Zander, Langstroth, Dadant, and Segeberger (foam hive). However, the most commonly hives being used in our country are Langstroth and Zander hives.

- *Langstroth hive*: It accommodates ten frames of 48cm × 23.2cm at 34.9mm centre to centre spacing (where 34.9mm is reduced to 31.8mm spacing it accommodates 11 frames).
- *Zander hive* is the most commonly used type and accommodates 10 frames.

These hives unlike the Kenyan top bar and fixed hive, is made up of detachable components namely, the hive cover, the inner cover, the super chamber (honey chamber), the queen excluder, the brood chamber (hive body) and the floor board (Fig. 4.9 and Fig. 11).

In the super chamber or honey super and the brood box (hive bodies), there are some moveable frames that are fitted with wax foundations. The hive cover acts as the roof of the hive and is usually made of a metal sheet. The hive bodies that contain the brood nest may be separated from the honey supers (where the surplus honey is stored) with a queen excluder.

1) Components of improved hive

Hive stand: A stand used underneath the bottom board or landing board of hive raises hive above soil level preventing excess moisture & rot. May also provide bottom insulation in winter or protection from ants, etc. depending on design.

Bottom board serves as the floor of the bee hive and is supplied with various means of reducing or enlarging the entrance to the hive. It can be made by piece of wood 55.88cm × 41.28cm wide × 1.91cm thick by joining two wooden boards together and nailing them in position (Harlan, 2001).

Brood chamber (hive body): Where the queen lays the eggs and the baby bees are raised. It provides space for egg and brood.

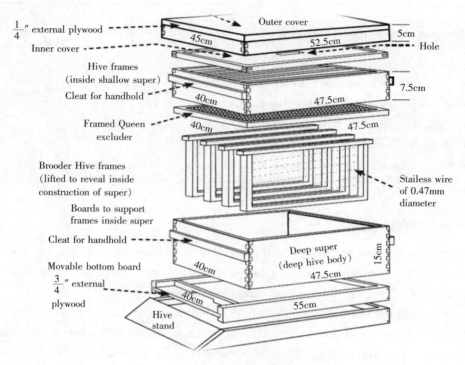

Fig. 4.9 Modern hive components
Sources: Ministry of Agriculture, 2012.

Supers refers as a medium where you'll be harvesting honey.

Inner cover: A rectangular coves and fits between top hive body and the telescoping cover or roof of the bee hive.

Outer cover is primary roof protects from rain, wind, snow and sun. Light outer covers need to be weighted down to not blow away in the wind (Brian Rowe, 2009).

2) Other hive components

Queen excluder is placed b/n the brood chamber and supers. Keeps queens out of honey supers and in the brood.

Wooden frame with foundation: Wood square frame uses metal wire to hold a wax foundation. The wax foundation will have a similar mold to the one piece frame.

Triangular escaped board removes bees from honey supers you want to harvest. The most stress free way to remove bees. The board is placed between the supers and the brood chamber. Bees navigate based on rules. When a bee reaches an obstruction, it will always travel to the right and follow that obstruction till its end. So, bees can leave through the 3 exits but not return.

Varroa Screen: It's made of a screen over a tray or sticky board. The tray is sufficient and superior as it catches the hive waste. When a mite falls into the tray through the screen, it will just sit there waiting for the next bee to come by.

2 EQUIPMENT FOR ROUTINE ACTIVITIES AND COLONY MANIPULATION

2.1 Materials for Foundation Sheet Making

Pure wax: Pure wax should be used (Fig. 12 A)

Frame wire: It is thin galvanized wire, which is stretched through the holes. It is used to support honey combs in the frame, i.e. prevent the curving down of combs due to weight and permits rapid handling and transporting long distance with little or no damage off (Fig. 12 B).

Casting mould: It is a metal coated with zinc. It is manually operated and used to make artificial comb foundation sheet (Fig. 12 C).

Embedded knifes is used as an alternative or hot iron bar to do the same purpose (as of transformer).

Frame: Wooden standard frame (Fig. 12 D)

Transformer: This transformer is used for fixing comb foundation sheets on the frame wires but it is not used in areas where electrification is lacking particularly in most places of rural Ethiopia (Fig. 12 E).

2.2 Feeds and Feeder

Types of feeders: The common types of feeders are top feeders, frame feeders, division board feeder, and board man feeder or external or entrance feeders. Top feeders and frame feeders are said to be indoor feeding method (Fig. 4.10).

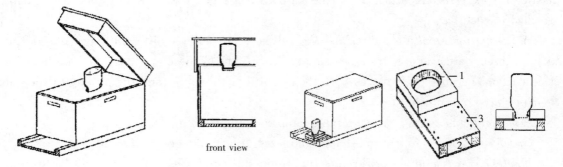

Fig. 4.10 Different types of feeders
1. Place for feeder 2. Entrance for bees 3. Sheet of zinc, tin, or plastic

> *Top feeders* are containers made up of either wood or plastic with a punched hole in the middle of the lid/cover placed inverted on the top of inner covers of hives.
> *Frame feeder* is a feeder suspended inside the hive in place of frame.
> *Board man feeders:* This is a feeder fixed or placed at the entrance of the hive. It is said to be out door feeding system.

> **Advantages of both frame & top feeders**
>
> Both weak and strong colonies have got their own share and an excellent method for feeding small amount of sugar syrup.
>
> **Disadvantages of both frame & top feeders**
>
> Labour intensive, i. e. required individual hive opening and linkage problem.
>
> **Advantage of outdoor feeding**
>
> Large number of colonies can be fed with minimum labour and do not require individual hive opining.
>
> **Disadvantages of outdoor feeding**
>
> It is liable to introduce robber bees, strong colonies are more benefited and the feed is exposed to spoilage.

2.3 Queen Rearing Equipment

The common equipment and tools required for queen rearing are:

Artificial queen cells can be plastic, wax or wooden.

A frame with a cell bar (s): The cell bar are made to fit in place of a frame and the plastic or wax cells are attached to the bar with hot beeswax, twenty cells are usually attached to each bar.

A 'grafting' tool (grafting needle): There are several different types of grafting tool, and most are available from bee-supply firms. The Chinese grafting tool is perhaps the easiest to use because it facilitates both the removal of the larvae and the placing of them in their prepared queen cells.

A magnifying glass: A magnifying lamp to help grafting

Queen cage comes in various shapes and sizes, plastic, wood or woven wire.

Nuclei box: A number of three to five frame nucleus boxes will be needed for starting cells and mating queens.

2.4 Farm Tools

Important farm tools used routinely for different activities such as land clearing and preparation for honey bee flora plantation, carrying materials in apiary includes: pick axe, sickle, machete, wheel barrow, spade and axe.

Machete: A large heavy knife with a broad blade used as a tool for cutting vegetation

Sickle: A curved blade used for cutting tall grass or vegetation

Axe: A tool with flat metal head with a sharpened edge used to chop/cut wood or tell trees particularly during swarm trapping

Pick axe: A tool used for breaking up or dig hard surfaces such as soil

Mattock: Tool like a pick axe with one end and forked at the other, used for loosening soil and cutting through roots

Shovel/spade: A tool with curved broad blade, used for lifting and moving loose materials/soil

Wheel barrow: A small cart used to transport materials, honey super and other things, usually in the form of an open container with a single wheel at the front and two handles at the back

Water sprayer: Used to spray water on bees to reduce aggressiveness and immediate evacuation from their nest, and also used during swarm trapping

2.5 Honey Harvesting and Processing Equipment

Honey harvesting and processing equipment include all equipment involved in opening of hive to remove honey crop from any kind of hive.

2.5.1 Honey Harvesting Equipment

1) Smokers

The body consists of a galvanized metallic sheet of gauge 28 canon and pumping bellow. The canon has a chamber with 2 holes, one for incoming air and the other one to let out smoke. Inside the chamber is placed a sieve to protect the inlet from being blocked with ash. The pumping bellow consists of 2 pieces of wood of size 12cm × 20cm, returnable spring, leather or canvas material. The construction dimension details are shown in Fig. 4.11.

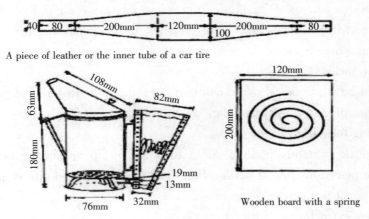

Fig. 4.11 Dimensional section of smoker
Source: Segeren, 2004.

2) Hive tool set

The hive tool set consists of a bee brush, hive opener and stainless steel knife (Fig. 13). The bee brush should be made of soft natural fibre, e.g. sisal fibre. One can also use bird quill feather or very soft leaves or grass, provided they are clean.

The hive opener is made out of flat iron bar of 6mm thickness, width of 25mm and length of 24cm. The hive opener is sharpened at both ends but curved at one end and should be painted with rustproof paint to avoid contamination of honey with corrosion and rusting with honey. The hive tool is a metal bar essential for prying apart frames in a brood chamber or honey super, separating hive bodies, and scraping away wax and propolis.

3) Hive box

Hive boxes and petri dishes are used to collect removed honey crop from hive until processing/extracting. It should be covered with same part of sheet/cloth to prevent robbing honey until arriving extracting room.

4) Weighing scale

It has a scaled clock face with a pointer, 2 hooks and re-setting nut. These weighing scales vary with maximum weighing capacity ranging from 25 to 200kg.

> **How to use a weighing scale**
>
> ➢ Re-set the pointer to 0' mark using a re-setting nut.
> ➢ Use upper hook for suspending the scale in a rope tied to a horizontal bar.
> ➢ Hang the container with the product on to the lower hook.
> ➢ Take the reading from the scale where the pointer ends and record.

2.5.2 Honey Processing Equipment

Important equipment used for honey processing includes: air-tight bucket, uncapping fork or knife, centrifugal extractor, honey settling tank, honey presser, honey strainer, refractometer, and honey jar.

1) Air-tight buckets

They should be white or yellow in color and of food grade material of capacity not more than 25kg for ease of transportation (Fig. 14A).

2) Uncapping fork/knife

Uncapping fork is mainly used to decap the cells of ripened honey before the framed honey combs are placed in the extractor (Fig. 14B). Uncapping knife is also used for the same purpose, but it is electrically operated.

3) Centrifuge extractor

It is a machine used to extract honey from combs and framed combs (Fig. 14C). It comes readymade. Some are made of food grade plastic while others made of food grade stainless steel. They have extracting capacity ranging from 2 to 18 frames. The combs or frames are arranged either radials, triangular or rectangular in order to extract honey. There are manual extractors as well as the electrical ones. All types have a spout for draining the honey out of the tank. The bottom is convex inside to allow all the honey to drain. They are fitted on 3 stands. The main body is cylindrical. They have 2 transparent plastic covers.

4) Honey settling tank

There are 2 types: food grade plastic tanks and stainless steel tanks (Fig. 14D). They vary in capacity from 25, 50, 100, 200 and 400kg. It has a cover and a spout with a convex bottom inside. Some come with inbuilt honey strainer while others come with separate double strainers.

5) Honey presser

It is used to extract honey of hand pressing of the honey combs which are not framed (Fig. 14E), e. g. honeycomb harvested from traditional and transitional hives.

6) Honey strainer

It is a double course screen. It is used in the normal processing of honey freshly extracted from the comb to remove the bits of wax that flow out of the extractor with honey. All honey as it comes from the extractor and before it goes into bottles (jars) should be run through a strainer to remove sediments and wax cupping.

7) Refractometer

It is a machine used to determine the percentage of moisture in honey. It is imported readymade (Fig. 14F).

How to use a refractometer

Open the slide cover and put a drop of honey sample and cover. Hold against light, view from the eye piece and adjust accordingly until you get a dark border line; where it marks is the percentage of the water content of that sample. Then, clean the slide thereafter.

---

8) Honey jars

It is glass or plastic contain sample of honey for marketing and each contains 0.5 – 1kg (Fig. 14G).

2.6 Personal Protective Cloth

The purpose of protective clothing is to protect the beekeeper from bee stings. Protective clothing should adequately cover the beekeeper and be of a light color (Fig. 4.12). Bees are sensitive to color and often become aggressive when they see dark or bright colors. Therefore, these colors should be avoided. All protective clothing must be cleaned regularly with a brush and water to remove the smell of the stings. Smell of stings triggers aggressive behavior in bees.

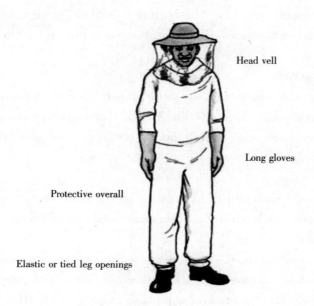

Fig. 4.12 Full personal protective equipment
Source: FiBL, 2011.

1) Head veil

The head veil protects the head and neck against bee stings. The head veil consists of a broad-brimmed round cotton hat which is under sewn with a black fine mesh measuring 25cm by 25cm and white gauze covering the rest of perimeter of the hat. The veil of the bee hat falls onto the shoulders and is tucked into the shirt or overalls. You should use black mesh or gauze for the window as it is easier to see through black mesh than white mesh.

2) Overalls

White overalls with a zip fastener must always be used for maximum protection. They must have elastic sown in the ends of the sleeves and legs, or the ends around the wrists and ankles are tied with elastic, rope or sticking tape.

3) Hand gloves and boots

Leather gloves and high shoes that cover the ankles must always be used. An extension piece of 20cm length with elastic in the ends is sewn onto the ends of the gloves. The shoes must cover the ankles.

>>> SELF-CHECK QUESTIONS

Part 1. Choice part.

1. _____ is part of hive components in which queen and brood found.
    A. Hive stand   B. Super
    C. Brood chamber   D. Bottom board

MODULE 4
IDENTIFY AND PREPARE BEEKEEPING EQUIPMENTS AND TOOLS

2. Among the following beekeeper's equipment, which one is used for removing extra combs and propolis from frames while hive opening?
   A. Smoker   B. Chisel   C. Brush   D. Forks
3. One is the advantage of modern hive over traditional (fixed hives)
   A. It can be open internally for inspection.
   B. The honey produced is mixed with wax.
   C. The hive produce more natural wax.
   D. Construction of hive requires less skilled manpower.

Part 2. Match the following equipment under column B with their function under column A.

A	B
1. Serve as the floor for beehives	A. Queen excluder
2. Rooms for eggs and broods	B. Hive body
3. Remove bees from super during harvesting	C. Smokers
4. Keeps queen out of super	D. Bee escape board
5. Serve as weapon for beekeepers during hive operation	E. Bottom board
	F. Chisel
	G. Honey chamber
	H. Frames

Part 3. Give short answer.

1. List the main beekeepers' equipment.
2. What are the types of beehives widely used in Ethiopia? Describe them according to their kind.

>>> REFERENCES

Cramp D, 2008. A practical manual of bee keeping [M]. Spring Hill House, United Kingdom.
Gebremichael B, Gebremedhin B, 2014. Adoption of improved box hive technology: Analysis of smallholder farmers in Northern Ethiopia [J]. International Journal of agriculture and extension, 2 (2): 77-82.
Rogala R, Syzmas B, 2004. Nutritional Value for Bees of Pollen Substitute Enriched with Syntheticamino acid [J]. Journal of Apicultural Science, 48 (1).

MODULE 5: ESTABLISH APIARY

>>> INTRODUCTION

An apiary is a place where beehives are kept. A good apiary management starts with choosing a good site to hang or place hives. The apiary should be clean to prevent disturbing insect ants from entering the hive. In order to establish apiary, it need availability of nectar sources plants like trees, shrubs and herbs. As nectar secretion is dependent on many factors (climate, weather, and soil), certain tree species, herbs and shrubs may not be good nectar producers when introduced into a new region. Check to see if the species of herbs, shrubs and trees are a good nectar producer under the conditions in the area where it will be growing, before advocating its use as a nectar source for bees (Haike, 2006). Therefore, the site must be in an area where there are several sources of nectar are available, meaning that it is preferable to place bees in the middle of the forage. The shorter the distance the bees have to fly, the less energy is lost and the higher the honey production (Segeren, 2004). This module is developed to provide you the necessary information on establishing an apiary.

1 OBTAIN SWARMS AND HIVES

Several methods exist of obtaining colonies for initiating beekeeping with honeybee species. Capturing swarms and buying honeybee colonies are method of obtaining colonies of honeybees. The easiest way to obtain honeybee colonies is obviously to buy complete hives from an established beekeeper. After obtaining colonies of honeybees, it should be transferred to standard hives properly equipped with frames and foundation (Pongthep, 1990).

1) Capturing swarms

Capturing swarms is one of the ways to obtain colonies for initiating beekeeping. Good baits and clean hives will help attract a swarm of bees to live in the hive. Hang the empty hive in the position that was occupied by the active hive (Pam, 2011).

2) Buying complete hives

The right price and the good condition of both the bees and the equipment, buying complete hives has often proved to be the most economical approach for beginners, who may in addition be able to obtain valuable suggestions and guidance from the seller. Other advantages are that an apiary can be established immediately and that the beekeeper can often divide colonies in the populous hives acquired. Considerations which the prospective purchaser should bear in mind in buying hives include the condition of the hive equipment, the population of adult workers and brood in each hive, the age and egg-laying performance of the queens, and the amount of honey and pollen stored, as well as the price.

3) Buying nucleus colonies

A nucleus colony is a small hive unit, normally consisting of 2 to 5 frames of brood, a small quantity of food reserves, several thousand workers and a laying queen. Nucleus colonies are cheaper than complete hives and are lighter in weight, so that they can be transported more easily at less cost. The guidelines set out above for the purchase of complete hives apply equally to the purchase of nucleus colonies. If possible, *nucleus colonies* should be bought in the spring, or at another time when natural forage is abundant.

2 SELECTING AN APIARY SITE

A general rule is that, the better the site, the better the colony will build up and the more honeybees' product you will be able to obtain. Everything you do now is aimed at obtaining a surplus of honey and other bee products at maximizing your harvest. Bees forage up to 3km, so apiaries should be situated within a minimum distance of 3km from bee forage places. The following important criteria should be considered to select good apiary site (Haike, 2006).

- Near water source but not near large rivers (as bees must cross the river to collect food)
- Bees prefer to work up hill.
- Bees prefer to work along rows of produced crops.
- Near good nectar forage places (forest, trees, nectar producing crops)
- Recommended distance from house 100m
- In shade, no direct sun, enough air circulation
- The character of soil in an apiary can have a direct effect on the bees especially during winter and rain period.
- Heavy clay or clay like soils is not desirable since they hold too much water.
- A sandy type of soil removes moisture immediately.
- A wet location with its resulting wet or damp bottom board can thus aid the spread of diseases (e.g. Nosema) and lower the brood rearing temperature in winter, this

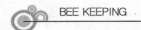

increases the humidity in the hive.
- High humidity can also aggravate a condition of dysentery.

3 CLEARING AND FENCING THE SITE

3.1 Clearing the Site

The apiary should be clean so that it is attractive and beautiful for the visitors and customers and any pests and predators are easily prevented and controlled.
- Tall grass and weeds need to be cut regularly.
- Branches of trees that reach the hives should be trimmed.
- To keep the grass and weeds away from the immediate front of the hive, salt is sometimes used for killing of all kinds of vegetation around the entrances; it must be liberally applied in front of every hive at the beginning of the season.
- Clearing obstacles and vegetation around hives allow easy movement; clearing the ground will also help to remove pests, such as beetles, toads and lizards.

3.2 Fencing the Site

Apiary is immediately necessary to enclose with the fence or hedge to prevent stray animals and vandals. Where bees are very defensive, planting a 2 meter high hedge around the apiary will force the bees to fly upwards over the heads of any people passing. If the plants in the live fence (hedge) are well chosen, they can provide food for the bees over the longest period of time, which will help to reduce absconding. Planting good bee plants can provide both food and shelter for the bees and, if carefully chosen multipurpose trees are used, can provide useful products for the household. If there is no natural protection from the wind to create around the apiary, you should plant windbreaks trees or tall shrubs as protective plantations (Pam, 2011).

In the forest areas, to accommodate the apiary you should first clear the area, freeing it from the tall trees, provided shade. Strong shading, especially on the east side, delays flights of bees in the morning. Similarly, the shadow in the afternoon causes premature termination of flights of bees. This can reduce the flight time of bee for 1 - 2 hours. Tall trees are uncomfortable by the fact that swarms go there that is very difficult to collect. Moderate shade from small and rare bushes, on the contrary, keeps hives from overheating in the hot midday hours, which ultimately increases the productivity of bees.

A fence is important for most backyard beekeepers. A six-foot high fence or shrubbery can serve several purposes: 1) forces bees' flight path above people's heads—bees normally travel in a straight path to their hive, a fence raises their flight path up over everyone's head, a fence reduces the chance that a bee will accidentally collide with someone walking nearby, 2) provides wind protection to the hives. Some people may be overly concerned about bees in the neighborhood. A fence hides most evidence that managed bees are in the neighborhood.

4 BAITING AND COLONIZATION OF THE HIVE

Beekeepers in many parts of the world depend on natural swarming to stock their hives, cavities or rafters. Rapid colonization of hives needs plenty of honeybee colonies in the area. Bees need to find the beekeepers' intended site and then to decide that it is the best place in the area for them to make their home. Positioning and baiting beehives in order to optimize their attractiveness to colonizing swarms will vary depending on the environment. Place the hives. Good baits and clean hives will help attract a swarm of bees to live in the hive. The best bait is beeswax because it smells good to the bees. Use plenty of wax around the inside of the hive and at the entrance (Pam, 2011).

In the first instance, whether it is natural nest sites, rafters or beehives that are to be colonized, if they have previously been colonized by bees then they will be more attractive than something that is new and raw. This is because the bees will be able to smell the residual odors of previous bees. The relatively fresh remains of dead bee brood and comb that has been used for raising brood is also very attractive. However, this also carries the danger of disease for the bees and it will quickly become damaged by wax moth cocoons, which will render it very unappealing so it cannot be used over a long period. Using a fresh starter strip of beeswax on the top-bars of a movable comb hive will act as a swarm attractant. Certain types of materials are more attractive to bees. It has long been noted that traditional hives are more quickly colonized than top-bar or frame hives. Plastic hives and other manufactured materials are often unattractive while some types of wood can have a strong smell, which is potentially repellent to bees. Scorched wood, where hives have been flamed to remove infection or pests, often seem to have additional interest, perhaps because of the minerals that may become available to scouting bees.

4.1 Materials for Baiting Swarm

Some indigenous herbs are used as attractants, in particular those smelling of lemon such as lemon grass. Among many other attractants people have tried are palm wine, banana skins and cassava flour—which may or may not attract bees but will certainly attract ants unless the hive is carefully set up. Indigenous knowledge is one of the greatest assets to a new beekeeper as it is local beekeepers have the most relevant practical experience to share. The size of the cavity, swarm catcher or hive also has a bearing on its attractiveness. The optimum size will vary with the size of the bee—with smaller honey bee ecotypes such as *Apis cerana* or African *Apis mellifera* being attracted to smaller sized cavities or hives. Bees have been shown to have preferences about the orientation of the entrance to the sun and whether the hive is in a shady position—temperate bees will avoid a shady place while tropical bees require it. An effective means of swarm catching is to use a special swarm catcher box placed along a known route for swarming or migrating bees. These routes are best identified by observation or by discussions with local beekeepers. Once colonized the swarm catchers can be moved to the main apiary. If a movable comb or frame hive is used, only the combs need to be transferred while the swarm catcher can be reused to collect another swarm. Other materials for baiting hive are:

- Little raw beeswax
- Dry cassava flour
- A sweet syrup such as palm wine or molasses, granulated sugar, sweet-scented lavender
- Limes, cow-dung, intestinal waste, lemon grass or even in very dry areas, a dish of water

4.2 Procedure to Baiting Swarm

1) Baiting a swarm only in useful swarming season, which is often in spring or at the beginning of the dry season

2) Taking a small hive that has already been inhabited by bees and use some indigenous herbs that used as attractants, in particular those smelling of lemon such as lemon grass

3) In the case of a movable beehive, fill this with frames or top bars.

4) Two of the frames should contain combs, the others should have foundation sheets or strips of old comb.

5) Placing hives along the swarming routes of bees

6) Placing the hive on a tree or on a roof in such a way that there is some protection from the wind is more attractive for swarm

7) Checking whether the hive has already been occupied by swarm

8) Moving the hive back to the desired place

9) Transferring the collected swarm to the normal hives

5 INSTALL BEE HIVE AND HIVE STAND

Let us now assume that the beehives and the site have been acquired. The hives must be installed, but before that is done, they must be prepared so that bees will occupy them. Clean the beehive. Be sure it contains no dirt, cobwebs, spiders or insect, which might arrest any scout bee visiting the installed beehive in the near future. Install hive on hive stands or hang hives.

In many areas, ants are a great threat to bees, and care must be taken to protect the bees from them. One solution is to suspend the hives with wires. A second solution is to use hive stands, which are protected from ants and termites by grease. Fix a collar of zinc or aluminum to each leg of the stand and grease the underside of these collars once a year with old crankcase oil. Make sure that no weeds, which could form a bridge for the ants, grow around the bee stands. As weeding can upset the bees, it is better to take the necessary precautions when preparing the apiary. You can also place old rubber mats or linoleum under the hive stands (Segeren, 2004).

A hive can be suspended, for example, between two trees or from sturdy branches of big trees. It can also be installed on a platform hive stand. This is a decision that must be made by the individual beekeeper. To avoid dangers to dampness and rotting of the bottom board, it is advisable to set it on pieces of boards, bricks concrete etc. When hives face the same direction in straight rows, the bees are apt to become confused at the entrances. Hives are arranged in pairs in such a way that they face each other with entrances 6m apart. When hives face south, the colonies get the benefit of a morning and a late afternoon sun, when the temperature begins to drop down. You should likewise have the most recommended beehives in your apiary in order to obtain a more yield (Gokwe south rural district, 2011).

It is important to make the hives with great care, so that there is (other than the flight entrance) no hole through which robber bees can enter. Robbers create unrest and lower the production of honey by the colony. All the hives and parts of the hives must have exactly the same measurements so that frames can be exchanged between hives. The outside of the hive should be protected with varnish or paint (preferably white). To reduce the chance that bees mistake their hives when they fly home, especially if the hives are neatly placed in a row, you can paint different geometrical figures 10 to 20cm high above the flight entrance of the hives or paint the flight board in different colors (Segeren, 2004). If the hive is to be inhabited by a small colony, you should temporarily reduce the size of the flight entrance by putting a stick in front of part of it.

5.1 Hive Hanging

Hang hives using strong greased galvanized wires to protect the bees from pests. Hang it in or under well-shaded trees. Suspend hives from wires so that predators such as the honey badger cannot push them over. Remember always when hanging hives that it is important to allow for ease of harvesting. Honey quality is improved by careful harvesting which is easier when the hive is within easy and comfortable reach. Use trees or solid poles to hang the hive. The hives should be hung at waist height above the ground. This is important in modern beekeeping as the beekeeper wears a bee suit making climbing difficult. Traditional hives are usually hung in trees. Alternatively, a hive can be suspended on a rope with a pulley that can be lowered for harvesting (Kangave et al., 2012).

Advantages of hanging beehives

- It is cheaper to hang a beehive than to install it on a platform.
- The lizard, an important hive predator, does not seem to pose a serious danger.
- Cattle and other grazing animals cannot tip the hive over.
- Running water cannot carry the beehive away.
- It is easier to prevent ants from reaching the hive than when it is installed on a stand.
- A thief seldom steals a Kenyan top-bar hive in a tree, especially when it contains honey, because it is not easy to remove the suspension wires if they are properly attached.

Disadvantages of hanging beehives

- A suspended hive can swing. The bees become alert and are prepared to pounce on the beekeeper if they find him.
- Honey-harvesting and brood-nest controlling are difficult to execute during the day.
- It is not easy to change the location of the hive. When removing it from the tree, the least false movement may result in tipping it over and jarring the whole contents. Sometimes the only way to remove it from the tree is to cut the suspension wires.

5.2 Install Hive Stand

Beehives should not be set directly on the ground. The main reason is that damp will get into the hive, and this must not be allowed to happen. A hive stand, therefore, is anything that keeps the hive off the ground. Stands can be pallets (four hives to a pallet), concrete blocks, bricks, wooden rails or simple wooden stands that hold one hive. In the main it is far better to improvise than to buy a stand from a beekeeping supply company, it is better in that it is cheaper and just as effective (David, 2008).

Bees evolved to live in trees not hives. A proper hive stand with a screened bottom board allows bees to be more hygienic as in tree cavities. In addition, a hive stand should be functional and nice to look at. The perfect hive stand should do be provided sturdy support for a hive off the moist ground. Recall that bees evolved to live in trees-not in the ground like yellow jackets. Yet the hive should not be so high that the beekeeper has to get on a ladder to reach the top supers. For a look at a hive stand that may be a wee bit tall, consider the beehive at the White House. Hive stand discourage animals, such as skunks or raccoons, which might be attracted by the protein-rich bees. Bees defend their hive by attacking the soft underbelly of skunks and raccoons. That belly is more exposed if the hive is on a stand. Hive stand also enable screened bottom boards to do their job-drop mites, hive beetles, and other junk out of the hive where they are less likely to find their way back. If you have ground under the hive, you can douse it with beneficial nematodes that will interrupt the small hive beetle's life cycle. It also keeps your bees relaxed and discourages ants. If ants are a problem, put your hive on a stand and put each leg of the stand in a bowl of water.

Advantages of installing a hive on a stand

- It is easier to place the hive on the stand and remove it.
- It is easy to move both hive and stand to another spot.
- The beehive does not swing about even if the beekeeper is working.
- Honey collection and brood-nest control can easily be carried out even during the warmest time of the day.

Disadvantages of installing a hive on a stand

- Grazing animals can knock the hive over.
- The legs of the stand can easily be used by lizards to reach the hive unless they are protected by lizard guard.
- It is more expensive and tedious to make a reliable stand than to buy a metallic wire for hanging a hive.
- Easy movement facilitates easy stealing. The thief has no time to waste.

Hive stands must be look like the following:

- At least one meter high
- It must be made of strong wood, so they will last long time.
- Hive stands will not rot: use types of wood that will grow easily into new plants when stuck in the ground.
- The legs of hive stands must be covered with grease or put in tins of oil so ants cannot climb into the hive (Fig. 5.1).

Procedures to install hive stand:
1) Clear the selected area/ground
2) Install the hive stand firmly long lasting poles, not rot quickly
3) Build the hive stand at a bout knee level
4) Do not arrange the hive stand in long straight rows, to reduce drifting bees
5) Making the hive stands should be protected from rot and damage by termites by using waste oil or grease on the legs. This will also prevent entry by ants (Fig. 5.2).

Fig. 5.1 Bee hive on tall stand
Source: www.beehacker.com.

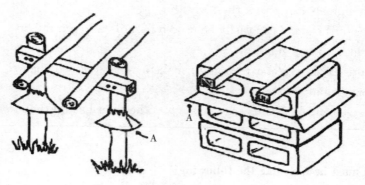

Fig. 5.2 Using greased caps (A) to protect bee stands from ants
Source: Segeren, 2004.

6 IDENTIFY HONEYBEE FLORA

African honey bees are not normally fed sugar. This makes it very important that plenty of nectar bearing flowers is available for as long as possible during the year. In many places, beekeepers and farmers plant multipurpose trees to meet their household needs. There is a great selection of multipurpose trees but trees that also produce nectar are very helpful for improving honey production. Try to select some plants that flower early or late in the season, so bees have more food during dearth periods. This will help to reduce absconding (Pam, 2011).

The greater the plant diversity, the more bees you will attract and support. Always try to choose as many native plants as possible, and consult with nursery staff or other experts to find vegetation that will thrive in your specific conditions. Honey bee friendly plants attract and nourish honeybees with nectar producing plants.

Honey and most other hive products do not originate directly with honeybees: they are natural products, which the bees have collected, and processed (Pongthep 1990). The bees visit flowering plants to obtain nectar, which is the source of honey, as well as pollen. Many plant species possess, inside their flowers near the base of the petals, glands called nectaries, which secrete nectar. Some plants have nectaries unconnected with their flowers, called extra floral nectaries. It should be recalled, however, that not all plant species have nectaries that secrete enough nectar to attract bees.

The concentration of sugar in nectar depends on several factors: the plant species and variety, the soil type, the time of day of collection, the temperature and relative humidity, etc. As a rule, plants with a higher sugar concentration in their nectar are more attractive to bees than that with weaker nectars. Because in the process of making honey the bees are obliged to get rid of excess water in the nectar, so that in treating more highly concentrated nectars, they need to expend less time and energy.

When the excess water has been evaporated from the nectar and the enzymatic reactions in the conversion of nectar to honey have been completed, the honey is ready for storage, to serve as the bees' reserve of carbohydrates to cover the colony's energy requirements. In the broad sense, then, honey is the colony's energy reserve, all or part of which will be expended in the process of foraging. From the standpoint of the beekeeper, a colony is productive when it stores a surplus of honey, i. e. when it can collect and convert into honey more nectar than they consumes. The beekeeper harvests all or most of this surplus honey. In some beekeeping systems, he/she may have to provide the bees with sugar syrup to replace the honey harvested, particularly at times when the colony requires additional food.

In both stationary and migratory beekeeping, the beekeeper seeks to place his colonies in or near areas where a sufficient quantity of honey plants—be they crop or pasture plants, weeds, shrubs, forest trees, roadside planting, etc. exists, in season or throughout the year, within the economical flight range of the foragers. Planting special crops for bees is not likely to yield a good economic return: arable land will provide better returns if it is used for other agricultural purposes.

6.1 Honey Plants and Pollen Plants

In order to survive, prosper and be productive, honeybee colonies, as has already been observed, must have a supply of both nectar and pollen in adequate quantities. Not all plant species are equally good for beekeeping. Some supply both nectar and pollen abundantly when in bloom and these are often called honey plants, because they are best suited for honey production. Plants producing nectar but little or no pollen are also considered honey plants. Other plants, however, may yield pollen but little or no nectar. These pollen plants are also important in beekeeping, especially at the time of colony build-up, when the bees need large amounts of the protein contained in pollen for their brood rearing.

Ideally, a good beekeeping area is one in which honey and pollen plants grow abundantly and with a relatively long blooming season. Such areas are however not always available or easy to find. The beekeeper therefore combines his skill in colony management with migratory practices in order to provide his bees with good, productive foraging environments. He/she must know the time and duration of the blossoming season of every major honey plant, including the environmental factors affecting them, and make a reasonable assessment of the supporting capacity of each area, i.e. the number of colonies that can be put to productive work there.

6.2 Floral Calendars

A floral calendar for beekeeping is a timetable that indicates to the beekeeper the approximate date and duration of the blossoming periods of the important honey and pollen plants in his area. The experienced beekeeper will have acquired much of this information over the years, but published charts are also available for many areas. The floral calendar is one of the most useful tools of the apicultural extension worker. It enables him to inform the beekeepers on what to expect in bee-forage availability, and when, so that they can manage their colonies in the most rational manner. Beekeeping in any specific area cannot develop without an understanding of the calendar, and for migratory beekeeping, special calendars for the different foraging zones along the migration route are required.

Assembling a floral calendar for any specific area is simple but time-consuming. It requires complete observation of the seasonal changes in the vegetation patterns and/or agro ecosystems of the area, the foraging behavior of the bees, and the manner in which the honeybee colonies interact with their floral environment. The accuracy of a floral calendar, and hence its practical value, depend solely on the careful recording of the beginning and end of the flowering season of the plants and how they affect the bees. The preparation of an accurate, detailed calendar will therefore often require several years of repeated recording and refinement of the information obtained. The steps normally taken in building up a floral calendar are as follows:

- The beekeeper makes a general survey of the area, drawing up a list of flowering plants found, special attention being paid to plants with a high floral population density per unit area or per tree.
- He places several strong honeybee colonies in the area, inspecting the hives regularly and observing changes for food stored within the hive to determine whether it is depleted, stable or increasing. Any food gains or losses can be monitored accurately by weighing the hives.
- At the same time that he monitors the hives' food stores, he surveys areas near the apiary and within the flight range of the bees, to record the species of plants that the

bees visit.
- ➢ He determines whether the plants are visited for nectar or for pollen. Pollen-foragers will have pollen pellets attached to their hind legs. To determine whether the bees visit flowers for nectar, the observer squeezes the abdomen of individual bees to obtain a drop of regurgitated nectar, tasting it for sweetness or measuring the nectar concentration with a hand refractometer.
- ➢ He studies the frequency with which the bees visit each flower species, in relation to changes in the level of the colonies' food stores. If there is a continuous increase in food stores, in direct response to the availability of the plants visited, the plants are good forage sources. When the food stores remain stable, the plants can be depended upon to meet the colonies' daily food requirements, but they cannot be classified as major honey sources.
- ➢ He carefully records all the changes in the blossoming of the plants visited. When the colonies begin to lose weight, the flowering season is finished for all practical purposes.

Once all the data on forage species have been assembled and repeatedly verified, they should be judged as they relate to the actual performance of the honeybee colonies. The calendar can then be drawn up in the form of circular or linear charts, showing the weekly or monthly availability of each plant and their flowering sequence.

6.3 Assessment of Areas for Honeybee Flora

Productive beekeeping depends on good colony management and good beekeeping areas, and in order to promote it as a profitable agricultural occupation, areas with a good potential for beekeeping must be located and evaluated. As in the assembling of floral calendars, weighing the hive is one of the most accurate ways of assessing the suitability and supporting capacity of an area. One major problem in this respect is how to select sites for assessment.

The following guidelines for the exploration and evaluation of potential beekeeping areas may be found useful:
- ➢ Referring to lists of known major honey plants in other countries or regions with similar vegetation patterns, agro-ecosystems, climate and edaphic conditions, determine whether similar plants are to be found in the area under study.
- ➢ The seasonal occurrence, in unusually high numbers, of bee nests can often indicate that there is ample forage in the area, at least during the period in question.
- ➢ The mere presence of flowering trees and shrubs in limited numbers, or of a few hectares of land covered with good honey plants preferred by bees, does not necessarily indicate that the area has potential for commercial beekeeping.
- ➢ Practical, large-scale beekeeping operations call for large areas, usually hundreds or thousands of hectares of nearby land bearing good forage with high population densities. Good honey plants are characterized by relatively long blossoming periods,

generally in terms of several weeks or months; high density of nectar-secreting flowers per plant or unit area; good nectar quality with high sugar concentrations; and good accessibility of the nectaries to the bees. The foraging land should be well proportioned, in terms of length and width, to promote foraging efficiency.

- ➢ The supporting capacity of an area for honey production is best determined by monitoring weight changes in the bee colonies. Among other factors that affect the economic value of an area for beekeeping are average hive yields, prevailing honey prices in the area, as well as costs of colony-management inputs.
- ➢ The fact that a flower is brightly colored or that it has a strong scent does not always indicate that it is good for bees, unless the fact is confirmed by the criteria set out above.
- ➢ The large-scale planting of honeybee forages has never been proved a profitable approach in terms of net economic return, except in integration with other agricultural activities, such as reforestation, roadside plantings, animal pasture, etc.

As nectar secretion is dependent on many factors (climate, weather, and soil), certain tree species may not be good nectar producers when introduced into a new region. Check to see if a tree species is a good nectar producer under the conditions in the area where it will be growing, before advocating its use as a nectar source for bees. Rainfall, temperature and sunlight affect the plants and thus determine the actual nectar flow. Weather also has an effect on quality. High rainfall promotes nectar secretion, but such nectar is often very low in sugar content. Conditions promoting optimum nectar flow are adequate rainfall before flowering and dry, sunny conditions during the flowering period.

Some good forage trees are listed (Haike, 2006): *Calistemon citrinus*, *Gemelina arborea*, *Ligustrum lucidum*, *Acacia* spp., *Croton megalocarpus*, *Grevillea robusta*, *Malus* spp., *Albera caffra*, *Calodendrum capense*, *Guazuma ulmifolia*, *Melaleuca* spp., *Albizia lebbek*, *Ceratonia silique*, *Erythrina* spp., *Prosopis juliflora*, *Azadirachta indica*, *Eucalyptus camaldulensis*, *Prunus serotine*, *Bauhinia variegate*, *Eucalyptus citriodora*, *Kigelia africana*, *Rhizophora* spp., *Burchillia bubaline*, *Eucalyptus globutus*, *Langunaria pattersonii*, *Syzygium cumini*, *Gliricidia sepium*, and *Leucaena leucocephala* (Fig. 15).

6.4 Seasonal Blooming Honey Plants for Bee Forage

Different honey bee plants bloom in different season. They are the very early blooming plants and trees that appear in spring. Practically none of these plants produces surplus, but many of them are valuable because of the fact that they enable the colonies to increase in strength in time for the later flowers. The earliest blooming plants furnish practically no nectar but they do furnish what is quite as important to the bee, and that is pollen. Summer time bee food comes in the form of many honey plants but not as many as found during spring. New England aster, goldenrod, spotted knapweed and Joe Pye weed from summer will continue to bloom into autumn or fall and feed your bees. Sometimes, earlier in the season, there is not much nectar coming from the goldenrod, and bees will choose other honey plants to get the nectar. However, in the fall, the goldenrod is ripe for the picking and the bees will work the goldenrod more during late summer and the fall, than the earlier periods.

Table 5.1 Some seasonal blooming honey plants for bee forage

Autumn	Spring	Summer	
Buckwheat	Elm	Clovers	Basswood
Spider flower	Maple	Alfalfa	Catnip
Sunflowers	Dandelion	Wild sweet clover	Horsemint
Fireweed	Hawthorn	Raspberry	Mustard
Smartweed	Red bud	Bee balm	Sage
Milkweed	Fruit trees	Blueberry	Sumac
Aster	Shadbush	Chestnut	
Goldenrod	Tulip tree	Corn	
Rape	Willow	Fig-worth	
	Sorrel	Locust	

Source: www.countryfarm-lifestyles.com/honey-plants.html

6.4.1 Common Honeybee Flora in Ethiopia

Beekeeping is one of the most important farming activities in Ethiopia (Workneh et al., 2008). It is one of the major incomes generating agricultural activity for the poorest and other beekeepers dwelling in areas where other livestock cannot exist and other income generating activity options are very limited. It also serves as means of income diversification for beekeepers in potential areas where other agriculture could be practiced including in rural and urban areas. Beekeeping contributes to country's economy through export earnings. Moreover, apiculture stabilizes and protects fragile environment and increase the production of agricultural food and cash crops through pollination service from honeybees. However, the success of beekeeping primarily depends on the availability of prevalent bee forages that based on its population density, nectar and pollen potentiality and prolonged flowering periods.

Ethiopia is endowed with diversity of plant habitat, climate, altitude and rainfall, as result quite large numbers of bee colonies exist in the country. Because of these diversity of plant habitat and environmental condition, distribution, flowering season vary from region to region. After the heavy rains between July and September, known as krempt, vast areas of the highlands are colored by golden-yellow because of the abundance of *Bidens* species Adey ababa, indigenous oil crops like *Guizotia abyssinica*, Nug, and red-violet with many clovers. Along path, forest edges, riverbanks and swampy meadows, a multiple of herbs are found growing and flowering, many of them not found elsewhere in the world: that is, they are endemic. Within regional and national forest priority areas, farmers are being encouraged to put out private domestic bee colonies, particularly along the edges of forests where bees can profit from flowering trees as well as the accompanying herbs and shrubs. Apiculture is deeply rooted in Ethiopia rural life and is basic to many cultural activities. Traditional knowledge by Ethiopian beekeepers and others of their botanical surroundings is largely unrecorded. This section of the flora of Ethiopia draws attention to some of the herbs, shrubs and trees that important to Ethiopian beekeepers (Reinhard and Admasu, 1994).

Table 5.2 Some common herbs and shrubs of honeybee flora of Ethiopia

Scientific name	Local name	Scientific name	Local name
Acanthus sennii	Kosheshla	*Lantana camera*	Yewof-qolo
Achyranthes aspera	Attuch	*Lathyrus sativus*	Gwayya
Acmella caulirhiza	Dame	*Launea cornuta*	Yeseytan gomen
Agave sisalana	Kacha	*Leontis ocymifolia*	Ras kmr
Ageratum conyzoides	Arema	*Lepidium sativum*	Feto
Ajuga integrifolia	Armagusa	*Linum usitatissimum*	Telba
Allium cepa	Qey-shnkurt	*Lycopersicon esculentum*	Timatim
Aloe berhana	Iret	*Maesa lanceolata*	Yeregna-qolo
Andropogon abyssinicum	Cajja	*Malva verticillata*	Adguar
Anethum foeniculum	Kamun	*Maytenus gracilipes*	Atat
Anethum graveolens	Selan	*Medicago polymorpha*	Wajema
Argemone Mexicana	Medafe	*Microglossa pyrifolia*	Hareg
Argyrolobium ramosissimum	Gerengere	*Morus alba*	Yeferenji'njorie
Arisaema enneaphyllum	Yelregnoch tila	*Nepeta azurea*	Damaka
Asparagus africanus	Kestennicha	*Nicandra physalodes*	Ate-faris
Becium grandiflorum		*Nicotiana tabacum*	Tmbaho
Berberis holstii	Zenqila	*Ocimum basilicum*	Beso-bla
Bidens macroptera	Adey ababa	*Ocimum lamiifolium*	Dama-kesie
Bidens pachyloma	Yemeskel ababa	*Origanum majorana*	Hassab
Brassica carinata	Yeguraghe gommen	*Osyris quadripartita*	Queret
Brassica nigra	Senafch	*Otostegia tomentosa*	Yeferes-zeng
Brucea antidysenterica	Waginos	*Oxalis corniculata*	Yebere-chew
Caesalpina decapetala	Yeferenj-ktktta	*Oxygonum sinuatum*	Kirnchit
Caesalpina spinosa	Qontr	*Parochaetus communis*	Yemdr-koso
Calotropis procera	Tobbiya	*Pavonia urens*	Ablalat

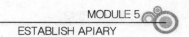

(continued)

Scientific name	Local name	Scientific name	Local name
Canarina abyssinica	Tutu	*Pentas schimperiana*	Weyinagift
Capsicum annuum	Qarya	*Persicaria nepalense*	Yetja-sga
Carduus nyassanus	Kosheshla	*Physalis peruviana*	Awtt
Carissa edulis	Agam	*Phytolacca dodecandra*	Ndod
Carthamus tinctorius	Yahya-suf	*Pisum sativum*	Ater
Catha edulis	Chat	*Pittosporum abyssinicum*	Lola
Caylusea abyssinica	Yerenchi	*Plantago lanceolata*	Goetoeb
Celosia argentea	Belbelto	*Plectranthus barbatus*	Yemaryam-weha-qeji
Cicer arietinum	Shmbra	*Plectranthus punctatus*	Yeoromo-dnch
Clematis hirsute	Azo-hareg	*Pterocephalus frutescens*	Henserase
Commelina benghalensis	Yewha-anqur	*Pterolobium stellatum*	Qent'affa
Companula edulis	Yeregna-msa	*Punica granatum*	Rooman
Coriandrum sativum	Dmblal	*Pycnostachys abyssinica*	Tontan
Craterostigma plantagineum	Babun	*Ritchiea albersii*	Dngay-seber
Crepis rueppellii	Yefyel-wetet	*Rosa abyssinica*	Qega
Cucumis sativus	Kiyare	*Rosa x richardii*	Ts'gie-reda
Cucurbita pepo	Duba	*Rosmarinus officinalis*	Yetbs-qtel
Cyanotis barbata	Yejb-dnch	*Rubus apetalus*	Njorie
Cyphomandra betacea	Ambarut	*Rumex nervosus*	Imbwach'o
Cyphostemma adenocaule	Asserkush	*Ruta chalepensis*	Tiena-addam
Datura stramonium	Astenager	*Salvia nilotica*	Basobila
Delphinium dasycaulon	Gedel-amuq	*Satureja abyssinica*	Sassag wucharia
Dichrostachys cinera	Ader	*Satureja punctata*	Tosgn
Discopodium penninervium	Aluma	*Scadoxus multiflorus*	Yedjib ageda
Echinops giganteus	Shok	*Scorpiurus muricatus*	Yebeg-lat
Eleusine floccifolia	Shinkur-ageda	*Sesamum indicum*	Mencha
Epilobium hirsutum	Yelam-chew	*Sida schimperiana*	Gorjejit
Ferula communis	Nslal	*Sideroxylon oxyacantha*	Tiffe
Galineria saxifraga	Solie	*Solanecio gigas*	Yeskkoko gomen
Galinsoga parviflora	Yeshewa-arem	*Solanum giganteum*	Mbwai
Glycina max	Akwri-ater	*Solanum melongena*	Berberjan
Gossypium hirsutum	T'ift'irrie	*Sorghum bicolor*	Mashla
Guizotia scabra	Mech	*Sparmannia ricinocarpa*	Chimki
Gynandropsis gynandra	Abeteyo	*Steganotaeniaceae*	Shunkwori
Haplocarpha schimperi	Getin	*Tagetes minuta*	Yahiya shito

(continued)

Scientific name	Local name	Scientific name	Local name
Helinus mystacinus	Galima	*Tapinanthus globiferus*	Tekatilla
Hygrophila auriculata	Yesiet-mlas	*Tirumfetta pilosa*	Sciamhegit
Hypericum quartinianum	Amja	*Tracnyspermum ammi*	Azmud
Hypoester triflora	Tqur-tleng	*Tribulus terrestris*	Aqaqma
Impatiens rothii	Gursht	*Trichodesma zeylanicum*	Koskus
Ipomoea tenuirostris	Yait-areg	*Trifolium acaule*	Maget
Isodon schimperi	Yefyel-gomen	*Trifolium burchellianum*	Alma
Jasminum abyssinicum	Wembelel	*Trifolium polystachyum*	Shal
Justitia ladanoides	Chingerch or Telenge	*Trigonella foenum-graecum*	Absh
Justitia schmpererana	Sensel	*Tylosema fassoglensis*	Yejb-ater
Kalanchoe densiflora	Ndahulla	*Urtica simensis*	Samma
Kniphofia foliosa	Ashenda	*Verbascum sinaiticum*	Dabakadet
Lablab purpureus	Yeamora-gwya	*Vernonia leopoldii*	Chibo
Lagenaria abyssinica	Kul	*Vicia faba*	Baqiela
Laggera pterodonata	Kes-kesa	*Zantedeschia aethiopica*	Yetrumba ababa

Source: Reinhard and Admasu, 1994.

Table 5.3 Some common trees of honeybee flora of Ethiopia

Scientific name	Local name	Scientific name	Local name
Dracaena steudneri	Chowyeh or Tabatos	*Ricinus communis*	Gulo
Mangifera indica	Mango	*Sapium ellipticum*	Arboche
Rhus glutinosa	Qmmo	*Dovyalis caffra*	Koshm
Schinus molle	Tqur-berberie	*Hypericum revolutum*	Amja
Cussonia holstii	Duduna	*Apodytes dimidiate*	Donga
Polyscias fulva	Yeznjero-wenber	*Persea americana*	Avocado
Schefflera abyssinica	Keteme or Qustya	*Acacia abyssinica*	Bazra-grar
Phoenix reclinata	Zmbaba	*Acacia albida*	Gerbi
Vernonia amygdalina	Grawa	*Acacia lahal*	Tkurgrar
Balanites aegyptiaca	Qacona	*Acacia pilispina*	Acq-grar
Jacaranda mimosifolia	Yetebmenja-zaf	*Acacia polyacantha*	Gmarda
Stereoespermum kunthianum	Washnt	*Acacia senegal*	Sbansa-grar
Adansonia digitata	Bamba	*Albzia gummifera*	Sesa
Ceiba pentandra	Yetit-zaf or Yekbritt-nchat	*Albizia lebbek*	Lebbek
Cordia africana	Wanza	*Calpurnia aurea*	Dgtta
Ehretia cymosa	Game	*Erythrina abyssinica*	Qwara
Buddleja polystachya	Amfar	*Leucaena leucocephala*	Lukina

(Continued)

Scientific name	Local name	Scientific name	Local name
Nuxia congesta	Chocho	*Milletia terruginea*	Brbrra
Boswellia papyrifera	Yetan-zaf	*Piliostogma thonningii*	Yeqolla-wanza
Opuntia ficus-idica	Qulqwal	*Pithecellobium dulce*	Nchet
Crateva adansonii	Dinkia-sebber	*Eucalyptus camaldulensis*	Qey-barzaf
Carica papaya	Papaya	*Diospyros mespeliformis*	Btremuseh
Maytenus obscura	Atat	*Erica arborea*	Asta
Maytenus senegalensis	Qoqqoba	*Croton macrostachys*	Bsanna
Combretum molle	Agalo		

Source: Reinhard and Admasu, 1994.

6.4.2 Toxic Plants for Bees

Plants can produce chemicals in sap, pollen, nectar or honeydew that are toxic to honey bees and humans. When environmental conditions, especially soil moisture, reduce other sources of nectar, the bees forced to forage from the toxic source because it is the only food available. Therefore certain bee plants are not good for honey making as the honey it produces can cause severe illness resulting in abdominal pains, nausea, headaches and even vomiting. This honey is known as poisonous honey. Plants such as rhododendrons, azaleas, and monk's hood all contain a glucoside of andromedotoxin. The following top ten plants are bad for bees (www.countryfarm-lifestyles.com/honey-plants.html).

1) Rhododendron

Spectacular and beautiful, not many people know the common *Rhododendron* hides a poisonous secret-its nectar is toxic to bees. Its common practice for beekeepers to keep their hives closed until the flowering season is over. The resulting honey from *Rhododendron* has contaminated honey, making it unsafe for humans to eat.

Alternative: *Clematis* has beautiful, wide flowers and is 100 percent bee-friendly.

2) Azalea

Rhododendron's sister, azaleas are also toxic to bees.

Alternative: Foxgloves are a bee favorite and despite being poisonous if consumed by humans, they are both honey and bee safe.

3) Trumpet flower or angel's trumpet

Though ornamental and sweet smelling, the trumpet flower's nectar can cause brood death in bees and is best avoided.

Alternative: Try honeysuckle instead for deliciously scented results.

4) Oleander

Harmful to butterflies as well as bees, oleander has a severe effect on hives. Nectar taken to the hive concentrates as it dries out, which increases the amount of toxins and usually results in a mass hive wipeout.

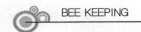

Alternative: Snapdragons are equally as bright and arguably more attractive in small or large gardens.

5) Yellow jessamine

Pleasantly aromatic and attractive as they are, the vines often poison bees and flowers of the yellow jessamine and its toxins are said to be as severe as hemlock.

Alternative: Plant black-eyed Susan in tubs and along fences for a pretty, easy to grow substitute.

6) Mountain laurel

Part of the blueberry family, the mountain laurel is an evergreen shrub with sweet, white or pink flowers when in bloom. Pretty they may be, but the honey produced by mountain laurel is toxic to humans and is often bitter tasting.

Alternative: Lilacs are both beautiful and wonderfully sweet smelling, easy to grow and are loved by bees and butterflies.

7) Stargazer lily

Stunning but deadly, stargazer lilies' pollen is poisonous to bees.

Alternative: Hollyhocks are impressive and just as beautiful as the stargazer but bee-friendly.

8) *Heliconia*

Exotic and interesting, *Heliconia*, or lobster-claws as it's sometimes called, is very toxic to bees.

Alternative: Although not quite as exotic, hyacinths are fragrant, gorgeous and easy to grow.

9) Bog rosemary

Not to be confused with the herb, bog rosemary is acutely poisonous and the honey made from this plant can cause paralysis to humans.

Alternative: Why not try planting a classic rosemary bush-aromatic, resilient and favored by bees?

10) Amaryllis

Now most commonly recognized as decorative Christmas flowers, *Amaryllis* is gorgeous in bloom but their pollen produces toxic honey.

Alternative: Dahlias are a highlight of late summer gardens. Beautiful and simple to grow, dahlias often flower until the first frosts of the year.

>>> SELF-CHECK QUESTIONS

Part 1. Choose the correct answer from the given alternative.

1. Which one of the following criteria used to select good apiary site?

    A. Water source   B. Near good nectar forage places

C. The character of soil   D. All
2. Which one of the following is/are used as attractants for bee swarm capturing?
   A. Lemon grass   B. Stargazer lily
   C. *Heliconia*   D. Bog rosemary
3. Which of the following indicate advantages of hanging beehives?
   A. It exposes bees for predator.
   B. Running water cannot carry the beehive away.
   C. It is easier to prevent ants from reaching the hive.
   D. B and C
4. Which of the following is/are the basic requirement to establish apiary?
   A. Finding the site   B. Obtaining colonies of honeybees
   C. Bee hives   D. All
5. Which of the following plant is/are toxic for honeybees?
   A. Dandelion   B. *Erica arborea*
   C. *Rhododendron*   D. Sumac
6. Which one of the following honeybee plants is/are found in Ethiopia?
   A. *Maytenus gracilipes*   B. *Commelina benghalensis*
   C. *Cyanotis barbata*   D. All
7. Which of the following materials is used for baiting swarm?
   A. Little raw beeswax   B. Dry cassava flour
   C. Sweet syrup   D. All
8. Which of the following is **not** advantages of installing a hive on a hive stand?
   A. It is easier to place the hive on the stand and remove it.
   B. It is not easy to move both hive and stand to another spot.
   C. The beehive is not swing about even if the beekeeper is working.
   D. Honey collection and brood-nest control can easily be carried out.
9. Which one of the following is **not** the importance of cleaning and fencing apiary?
   A. To make attractive and beautiful for the visitors
   B. To delays flights of bees
   C. To force the bees to fly upwards over the heads of any people
   D. To control any pests and predators
10. Which one of the following is/are considered in building up a floral calendar?
    A. The beekeeper makes a general survey of the area.
    B. Places several strong honeybee colonies in the area
    C. Determining whether the plants are visited for nectar or for pollen
    D. All

Part 2. Match the correct answer from column B to column A and write the letter you choice on the space provided.

BEE KEEPING

A	B
_____ 1. Time table for flowering bees plant	A. Bait hive
_____ 2. Keeps hive off ground	B. Poisonous honey
_____ 3. Prevent hive stand from ants	C. Waste oil or grease
_____ 4. Toxic for bees	D. Hive stand
_____ 5. Used for capturing swarm	E. Floral calendar
_____ 6. Protects person from bee sting	F. Goldenrod
_____ 7. Autumn blooming honeybee plant	G. Wild sweet clovers
_____ 8. Spring blooming honeybee plant	H. Azalea
_____ 9. Summer blooming honeybee plant	I. Personal protective equipment
_____ 10. Honey from *Rhododendrons*	J. Tulip tree
	K. Smoker
	L. Beehive

Part 3. Write the correct answer accordingly.

1. Define apiary.

2. Write the basic equipment, materials and tools required to establish apiary.

3. What are the criteria should be considered to select apiary sites?

4. Mention at least three advantages and disadvantages of installing a hive on hive stand.

5. List at least five top plants honeybees loves that flower in spring, summer and autumn.

6. Write at least ten shrubs and herbs of honey bee's flora that found in Ethiopia.

7. Write at least ten tree of honey bee's flora that found in Ethiopia.

>>> REFERENCES

Cramp D, 2008. A practical manual of bee keeping [M]. Spring Hill House, United Kingdom.

MODULE 6:
CARRY OUT ROUTINE BEE KEEPING ACTIVITY

>>> INTRODUCTION

The goal of honey bee colony management is to aid the colony to build up to its maximum during the main nectar flow and to survive the dearth. Well-managed colonies assure the greatest possible return for the beekeeper. The first management step in beekeeping is obtaining bees in a manageable hive. Once the hive is established, it should be inspected regularly and managed according to its need.

All the inputs necessary for carrying out a beekeeping venture can be made locally. Local tinsmiths, carpenters, or basket makers can make smokers, protective clothing, veils, and hives. Thus, beekeeping activities can create work and income for these people. Small beekeeping activities can be successful from the beginning. After a project is started and expertise is gained, it is easy for a beekeeper to increase the number of hives. A dependence on outside resources or inputs is not necessary to do this. Bees feed themselves from the existing nectar and pollen resources of the area by foraging far beyond the small amount of land on which the hives are located.

Traditional beekeeping is the oldest and richest practice, which has been carried out by people for thousands of years. In Ethiopia, beekeeping is one of the oldest agricultural activities having been passed from generation to generation without being modified up to present times. Several million-bee colonies are managed with the same old traditional beekeeping methods in almost all parts of the country. Beekeeping has been and still is very widespread and economically important for the farming community of the country. The farmers to produce their bee baskets use cheap local materials like clay, straw, bamboo, false banana leaves, bark of trees, logs and animal dung. Like in other branches of agriculture, they do not need to invest, and they use very few tools, mostly knife. Almost all methods are based on the concept of minimal management.

Moveable frame hives (Zander type) were introduced into Ethiopia in 1965 and many demonstration sites were established to observe the adaptation of these hives and the Holetta Bee Research Center was established for technological and management improvement of beekeeping development through introduction and dissemination of modern beehives and research results. However, the rate of penetration of modern hives and the progress of the subsector had been slow. This module enables to perform routine beekeeping activities.

1   HANDLING BEES

The beekeeper should take into account that bees react strongly to certain smells such as perspiration, alcohol, soap and perfume. In order not to be stung, avoid carrying these strong smells when you inspect the bee colonies and do not keep any animals near the bees. Bees can also become entangled in hair and in woollen clothing. It is therefore advisable to cover the head and to wear clothes made of smooth fabric. When bees are aggressive they will always go for dark colours first. Wear clothing of the lightest possible colour. This is also better when working in hot climates. Make sure that you always have some form of smoke at hand when you want to open the hives. Always first blow some smoke into the flight entrance, especially if you are working with the more defensive kinds of bees in Africa and South America. Lift the cover, blow some smoke into the hive and close the hive again for one minute. Always make sure that you have enough fuel for the smoker at hand.

Some types of bees are easily disturbed when vibrating objects, especially machines, come close to them. Avoid this by choosing the site of the apiary carefully. Weeding or mowing grass with a sickle or scythe can excite the bees terribly. Carry out all activities with slow movements. Bees react strongly to rapid movements. Even if you have been stung, first calmly put the frame back into the hive before paying attention to the stinger. You should especially avoid banging against the hive.

If you are stung, you must first kill the bee that has stung you and then scrape the sting out of your skin with a fingernail or a sharp object. When you first start keeping bees the stings will cause swelling. After several stings the reaction will become less. If you react violently to a bee sting by perspiration or dizziness, it is advisable to stop keeping bees. Fortunately this reaction only occurs in 1 out of about 5,000 people, but if it does occur, go to a doctor immediately.

2   MAINTAINING CLEAN HIVES AND APIARY

Hive and apiary maintenance: Bees are very clean and want a clean home. They will not move into a hive that is leaking or has rats, spiders or other undesirable creatures already living there. If a hive remains un-colonised after the swarming season, clean it out, sterilise over a fire and add new baits.

➢ Proper maintenance extends the life of the hive.
➢ Check apiary for hive conditions.
➢ Inspect for rotten, loose or broken boards and frames.

- Reconstruct, tighten or replace frame parts.
- Paint supers with light colors to beat summer heat.
- Take advantage of the winter months to do maintenance and prepare for the new season.
- Repair clothes, veil, gloves and bodysuit.
- Inspect your essential two pieces of equipment, smoker and the hive tool.
- Inspect and repair trucks, trailers, and loaders.
- Repair fences.
- Eliminate trash in the apiary.
- Practice fire safety when the bee smoker is in use.

Swarming control: Honeybee colonies should be managed to minimize swarming. Beekeepers who learn of a nearby swarm should take reasonable measures to see that swarms from their hives are retrieved to prevent it becoming a nuisance.

Provision of water: Beekeepers need to provide a suitable source of continuously available water for their bees.

Disease control: It is incumbent on beekeepers to monitor and manage disease and pests to ensure colony health. Beekeepers should take remedial action to prevent spread of disease.

3 ADMINISTRATION ACTIVITIES

You should keep a careful record of the condition of the colony, especially if you have several colonies. Make notes on a card after each inspection of: the date; the presence of brood combs; the food supply; whether there are drone or swarm cells; and also any action you have taken. Also note the honey yield or the absence of yield and any other particulars such as aggressiveness. You can attach the hive card to the underside of the cover of the hive.

Instead of using the card system, you could also write all the details in a notebook, or better still a loose-leaf file, which you take home with you. It is difficult to write on hive cards if you wear gloves when working with the bees, so scribble a few notes on a piece of paper and write them up in detail when you are home. For administrative purposes, it is useful to number your hives. All the data collected will be very useful at a later date when the number of colonies in the apiary has grown considerably and you want to start selecting the best ones.

Successful comb management is an important part of the beekeeping practice. Combs used for brood rearing change in different respects. The comb color turns from yellow to brown and black. The dark color of old combs is caused by larvae excrements, pupae skins and from propolis rests. The properties of the combs change too: cells become smaller and thicker. These changes result in the production of smaller bees. Apart from these changes, old combs are sources of infections. Honey, stored in dark combs will also get dark and dirt particles will contaminate it. Feed will also crystallize more readily in old combs, thus making hibernating more difficult. Old combs contain less wax and more protein and will be more readily attacked by the wax moth.

4 MOVING HIVED COLONY

For various reasons, you may have to move your hives to another apiary (for example, to avoid spray damage, to pollinate a certain crop or just because you are moving). Moving bees isn't difficult, but there are two major problems (Cramp, 2008).

If you move your bees within their radius of foraging, then, once you have moved the hive, the foragers will all fly back to the original position. This means you should move your bees at least 2 - 3km. Even if you move them 18m, they will fly back to their original position and cluster there. If you move them just a metre, however, they will usually suss things out and go to the right place.

The second problem is that, if you move your bees during the day, you will lose most of your foragers because they will be out. You can, however, block them in the night before with equipment similar to that used for spray protection (mesh entrance block and gauze lid), and then move the hives the next day. The bees will panic and suffer stress but, if you have provided them with room, water, food and ventilation, they should survive.

4.1 Long Distance Moving Hived Colony

Before moving your bees, you must ensure the hive boxes won't come apart. This includes the floor and the lid, unless the lids are telescopic. Use straps, staples between the boxes or clips of some kind. Metal straps are the best method. The basic rules for moving bees are, then, as follows:

- Always strap the hives up tight.
- Ensure the hives won't shift and are strapped down well.
- Move your bees more than 2 - 3km away so that you have no problems with your bees returning.
- Move your bees up to a metre away and your bees will have no problems in returning.
- At night, load up, strap down and go.
- In bad weather, load up, strap down and go.
- During the day, shut the hives the night before when all the bees are in, and allow plenty of room, ventilation and water.
- For short moves of up to an hour, block the hives up at dawn before the bees are out and then go.

4.2 Short Distance Moving Hived Colony

But what if you need to move your bees only 18m or so? This can be done but it is a little laborious. You could take 18 days to do this by moving a metre a day, or you could move the hive and place a dummy hive with comb on the original site to collect the foragers. Then, at the end of the day, you put the dummy hive onto the moved hive.

That evening, block the entrance with grass and, by the time the bees manage to chew their way out, they will have gathered that something is different and may then take new orientation flights. You may need to repeat this procedure until the foragers learn where they are meant to be.

5 WEIGHING HIVED COLONY

A beekeeper can assess the condition of a colony from outside the hive. The behavior of the bees, and the presence of drones, indicate whether the season is going up or down. The presence of a 'beard' of bees indicates that the hive is full. With experience, the beekeeper knows exactly the seasons in the local area. However, the number of combs within the colony at any time can be known only by internal inspection of the colony. There is a way to gain objective data without opening the hive, and this is to weigh it. This is an especially good option for top-bar hives, as they contain only natural combs.

Subtraction of the empty weight of the hive gives the weight of the colony—that is combs and bees. To weigh hives in the apiary, three people are needed: two to hold and lift the hive, and one person to shift the scale underneath and let it stand properly. The weight can then be read. Two readings must be taken to be sure. The operators have to get used to the mirror reading, while the hive is standing free. The first reading is noted down and then the second before the hive can be lifted again to remove the scale.

The hives are numbered. The empty weight of the hive is obtained beforehand. If hives are already colonised, it can be obtained later if the colony absconds or is transferred into another hive. The top-bars are laid in place, the lid put on and the weight is recorded in a table with other data, including the name of the beekeeper and where the apiary is located.

The hive does not need to be opened to have an idea how full it is. It is found that the maximum weight of bees and combs in kilograms is about half the volume in litres. This is at the peak of the honey flow when the colony has its maximum weight. For example, if the volume of the top-bar hive is about 80 litres, the maximum weight is 40kg. A hive of 50 litres contains a maximum of 25kg, and one of 100 litres a maximum of 50kg. Seasonal management of colonies can now be based on objective data. If weighing is done monthly, a graph can be made to give a good insight into colony development during all seasons.

6 MAKE COMB FOUNDATION SHEET AND FIX TO FRAME

The term used for a thin sheet of beeswax embossed with the hexagonal pattern of comb. In frame hive beekeeping, a sheet of foundation is placed in each frame to encourage the bees to draw out cells in the size and orientation required by the beekeeper. Movable frame hives allow drawn comb to be re-used several times, for brood or honey storage. Beekeepers may use wire in foundation to increase the strength of the comb. This consists of a sheet of beeswax (2 - 3mm thick) into which a cell pattern has been pressed. The bees build up the cell walls on this pattern (FAO, 1985). The use of foundation sheets promotes the construction of a regular comb, provided they are made of pure beeswax and the right cell size is pressed. Furthermore, the bees need to use less energy to produce wax, which is very advantageous for honey production. Foundation sheets are commonly used in frame and hive beekeeping, but are not strictly necessary. If they are not available, pieces of so-called burr comb (newly built small pieces of comb, usually in the shape of a tongue) can be used.

The beekeeper can attach burr combs to the centre of the frame bar or top bar by melting its top. If foundation sheets are available, but expensive, then it is more economic to use small strips of foundation sheets only. This type of foundation sheet can also be used in top bar hive beekeeping (P. Segeren, 2004). Combs are produced basically by two methods: by sheeting and by casting (milling). Sheeting of beeswax was the first method used in production of foundations. In a first stage wax sheets were produced and in a second the foundations were produced by calendering. The wax sheet was run through a foundation mill, which would print the foundations. Today, Foundations produced by sheeting and milling is the preferred method world-wide. Casting or milling of wax will produce foundations that are more brittle in the cold than milled sheets. Cast foundations are produced mainly by beekeepers, as this method is easy to perform in small beekeeping units.

The frame holds the sheet of foundation beeswax on which bees build their cells. The foundation is fixed to four horizontal strands of wire that keeps the foundation perfectly straight and vertical. The wires also help to strengthen the newly constructed comb. Combs well supported by wire seldom break in the extractor or during transport. Suitable galvanized and stainless steel frame wire is available from beekeeping equipment suppliers. Pre-wired frames are also available from some outlets. Frames have end bars already drilled for the wires. Methods commonly used to prevent the wire cutting into the grain of the wood when tightened and so losing tension are: placement of an eyelet in each wire hole, counter sink holes with a centre punch to harden the wood, placement of a staple on the tension side of each hole. Molten wax is poured into a tray between two embossed moulds (lightly wetted with mild soapy water to enable easy removal). Alternatively thin, flat sheets of beeswax may be made in simple moulds and then rolled between embossed formers to create the hexagonal cell pattern. Presses can be made of metal, plaster of pairs or plastic. The sheets of embossed wax are then cut into the rectangular sizes needed for frame hive beekeeping. Wired foundation can be made by fastening wires into the frames, then heating these with a low electric current to melt them into the sheet of beeswax foundation and the frame. The cell size required in foundation differs with species and race of bee. Most frame hives have been designed for European *Apis mellifera* and most foundation is made for this size of worker cell, usually 5.4mm. Smaller cell sizes, for example 4.7mm, are available in some countries. Larger drone cell size is also sold, primarily for Varroa control by means of removing drone brood (which is preferred for Varroa reproduction) from the colony.

Reasons for using foundation sheet

➢ In frame hive beekeeping, foundation encourages the bees to build comb in the frames that are of the size and orientation required by the beekeeper.
➢ Foundation enables bees to draw out comb more quickly, reducing the effort and resources spent in wax and comb production, allowing more time and space for honey production.
➢ Cells of standard worker size are encouraged, rather than drone comb. This will increase the number of worker bees in the colony and save resources used for rearing drones.
➢ The use of foundation and frames allows beekeepers to manipulate colonies, remove and replace combs, and move combs with or without bees, brood or stores, between hives.
➢ Foundation and frames allow beekeepers to do many interventions that are convenient for them, and provide an easy way to handle bees, for example to help beginners to get started with bees.

6.1 Materials Required for Making Foundation Sheet

Materials required for make foundation sheet contain beeswax, casting mould, detergent (omo), water, alcohol, pan, knife, sackcloth or a sack (preferably jute), string or twine (2 – 3 metres), a stick or a discarded top-bar, a large spoon or ladle, a mould for the wax, fire (heat source), and table.

6.2 Procedure in Making Foundation Sheet

1) Melt the pure wax, adding water in the pan.
2) Make the solution for smoothing (omo, honey, alcohol).
3) Wash the casting mould by the solution of the above listed.

4) Add the melt wax in the casting use spoon.
5) Cover the casting mould.
6) Add the water on casting mould.
7) Stay for certain minute.
8) Open the casting mould.
9) Trim the comb according to the size of frame.

6.3 Fixing Foundation Sheets into Frames

Comb foundation sheet can be fixed on the frame wire in different ways:
- *Using electric transformer (18-24 volts)*: Heat the wire by running a peak electric current thought it and the wire will sink in the foundation sheet.
- *Using spur embedder*: The spur embedder is heated in boiling water or over flames and run along the wires; it melts the wax sufficiently to allow the wire to sink in.
- *Using soldering iron (hot iron)*: Heat the iron with flame. Using soldering iron press the wire into the foundation sheet, the wax will melt and fix the foundation sheet with the frame wire.

Factors to be considered when we undertake fixing wax into frames:
- The foundation sheet is at least 0.5cm smaller than the frame at the sides and bottom.
- *Avoiding bending*: Foundation sheet must stick to the upper edge of the top bar or the frame (Fig. 16).

7 PREPARE SMOKER FOR USE

No honeybee will ever allow a beekeeper to harvest its honey without a fight. The tropical honeybee is noted for its aggressiveness, and the beekeeper is warned not to conduct any brood control or harvest without using his smoker. The smoker has two main parts: the container which is a metallic can, big enough to carry enough dry material to last at least 40 minutes; and the bellows section, which puffs air into the container to drive the smoke out of the can.

The container is loaded with wood shavings, cow-dung or any dry material which provides white smoke (no oil or kerosene should ever be used in a smoker). The smoke renders bees docile, so that the beekeeper can work undisturbed (Adjare, 1990). The body consists of a galvanized metallic sheet of gauge 28 canon and pumping bellow. The canon has a chamber with 2 holes: one for incoming air and the other one to let out smoke. Inside the chamber is placed a sieve to protect the inlet from being blocked with ash. The pumping bellow consists of 2 pieces of wood of size 12cm × 20cm, returnable spring, leather or canvas material.

Recommended materials for smoking contain semi dry grass, wood shaving, coffee husks, maize comb, bean husks, millet husks and dry cactus. Recommended procedures in lighting a smoker:

1) Put glowing charcoal on the sieve in the canon chamber, followed by any of the smoking material mentioned above, filling the chamber 3 quarter way.

2) Before closing the smoker, fill the canopy, i.e. the last quarter with fresh green grass or leaves. Gently, start pumping the bellow and smoking.

8 OPEN AND REASSEMBLE THE HIVE

The best time to open hives is in the middle of a warm day, especially when the bees are busily engaged in collecting nectar. Bees should never be handled at night or on wet cold day. It is not always possible for the extensive beekeeper to choose the ideal time but it is well to plan to open hives in favorable periods, not only for the comfort of the operator but principally to interfere least with the work of the colony. In practice, the aggressiveness of the African honeybee makes it impossible for most beekeepers and wild-honey tappers to approach their hives or harvest their honeycombs in broad daylight. Comb moving and most related jobs, such as brood-nest control, are best performed late in the afternoon or delayed until night or early morning, when bees are less aggressive. This explains why most honey tappers work at night. However, it is not easy to work well in the dark. Light must therefore be provided, and this definitely requires an extra hand to assist in the operations. Flashlights, which are ideal for use in the job, are usually beyond the reach of the average honey-tapper, especially since, in many tropical countries, batteries for them cannot be obtained on the market. When lanterns or hive torches are used, many bees, attracted to the fire, are burned to death. If the lanterns are shaded to avoid this, bees will cluster around the shades and shut out most of the light (Beekeeping in Africa, 1990).

The entire comb which is removed from the hive has to be returned in their order. Do not interchange position of combs if they are removed. Take care not to crush bees during the reassembling process. In closing a hive, after the frames are replaced and spaced properly, the cover should be put on in such a way as not to crush the bees. If necessary the bees may be driven down by the use of smoke, but if bees are on the top edges of the hive, the cover may be slide on from the end or slides so that none will be crushed.

9 INSPECT THE COLONY

Colony inspection is one of the colony management, which enables to monitor honeybee's activity and ensure that the maximum strength of the colony coincides with the maximum nectar flow and others in order to obtain a maximum honey production measures. Once the hive is occupied and the bees are busy, it is said to be colonized and it is important to inspect the colony to monitor its performance. There are two ways of colony inspection:

internal and external. The experienced beekeeper can usually have a fair idea of how his colonies are progressing by observing them from outside, the only means he has of knowing for sure whether everything is going smoothly is to open the hives and inspect each comb (Stephen, 1990). This will let him know if honey is being prepared and capped regularly, whether the colony is getting ready to swarm, whether the hive has been attacked by pests, etc.

One of the advantages of modern hive and transitional hive is the possibility of inspecting the hives internally to identify the problems and take a control there are the precautions to be followed during inspections.

During periods of food abundance in the spring and summer, drones will also be present in a healthy colony. In addition to the adult bees, a healthy colony will also have *brood*, the collective term for eggs, larvae, and pupae. Once a week take a quick look into the hive. The bees should not really be disturbed that often, but as a beginner you still have a lot to learn about the life of the bees and this will have to be done at their expense. Inspect the colonies during the day, when the weather is sunny, but preferably not when there is a thunder storm on the way.

Note the following points: Are there eggs, larvae, capped worker brood or drone brood? Is the queen present? Is there enough food? Are there any wax moth larvae? Are the bees and brood healthy? The findings are noted on the hive card. You must always hold the frames above the hive so that the queen does not fall outside the hive. To prevent diseases spreading, especially regarding American foulbrood, it is recommended to sterilize equipment prior to inspecting the hives of other apiaries. Gloves should be washed between use in one apiary and before use in another. Scorching the hive tools in a lighted smoker will also prevent the transfer of spores between the hives (Segeren, 1988, 1991, 1996, 2004).

9.1 General Guidelines to Inspect

- From colony to colony and from season to season
- When colonies are newly transferred or established
- During death period whether there is food or not
- After the first bloom, to check for growth, strength of the colony, signs of swarming, etc.
- After a major honey flow to add or remove supers
- To cheek condition of the queen & brood
- After queen introduction
- Whenever disease, queenless, pesticide damages, etc. are suspected. Do no inspect/examine hive: during peak/major honey flow unless necessary, e.g. if disease is suspected, to re-queen, or to add or take off supers, on a very windy, cold or warm day, when it is raining at night.

9.2 External Inspection

External inspection gives many clues about the status of the colony. During external inspection, observe the following points:

- Lay out of the apiary, including wind breaks, water sources, feed source etc.
- Weather conditions: wind, temperature, humidity
- The normal in & out movement of bees (watch number of bees coming in and out of the hive)
- The load carried in by the bees (types of food brought to the hive)
- The dead bees around the hive or at the hive entrance, unusual occurrence
- Accumulation and clustering of bees at the hive entrance
- Presence of bee enemies (ants, spiders)
- Weeds & other dirt around the hive
- Presence of drone around the hive

9.3 Internal Inspection

9.3.1 Requirements for Internal Hive Inspection

- Select time of inspection depending on the weather and site of the apiary, usually after noon.
- Optimum weather, not too cold and not too warm
- Careful dressing, proper wearing of protective clothing to avoid unnecessary stings
- Wash previously used protective clothes to avoid the alarm pheromones of the previous stings.
- Light clothes
- Use proper smoke and other necessary equipment to calm down the bees.
- Gentle movement, operate gently to avoid crushing of bees
- Always be behind or from sides of the hive.
- Start from weak colony.
- Do not open colonies one after the other (Fig. 17).

Inspection of honeybees must be done on possible shortest time

- To avoid much disturbing of the colony
- To avoid robbers coming from other hives
- To protect the brood from chilling
- To minimize excessive stings
- To minimize consumption of stored honey by bees

9.3.2 Internal Hive Inspection Procedure

The general rules for hive inspection and for harvesting honey are the same, and therefore they can be discussed together here.

1) Wear protective clothes, and cover the body thoroughly. It is better to have another person check to be sure the bees have no way to reach the skin.

2) Beekeepers should always work in pairs: one operating the smoker and the other working the top-bars and combs.

3) Get a good smoker with a large bellows. The fuel container must be large enough to carry enough fuel to last for the entire operation. Carpenter's wood shavings are excellent for fuel. Never forget to take along a good knife or hive tool and brush or quill.

4) Puff smoke into the entrance of the hive. Allow 10 to 20 seconds for the smoke to calm the bees before opening the hive. After opening the hive apply additional smoke by puffing it under the lid, between the supers and over the frames. Be careful not to 'over smoke' —if you over smoke, the bees will run to excess.

5) Using the hive tool or knife, pry open the lid of the hive if it has been propolized (top-bar hives have no problem with propolizing).

6) Then remove the first comb and inspect it. If it is a brood comb, look to see that the cells are filled regularly and well-sealed, and especially whether the comb contains queen and drone cells as well as worker cells; this is a sign that the colony is preparing to swarm. If it is a honey comb, look to see whether the cells are fully capped (containing ripe honey) or uncapped or partly capped (containing unripe honey). Then replace the comb, even if it is full of ripe honey; it can be removed and taken away later, during honey-harvesting operations, which call for special equipment.

7) Replace the comb; give a puff of smoke, go on to the next comb, and repeat the operation until all the combs have been inspected.

8) If more than ten brood combs are found, remove the excess, because if too much brood emerge, the hive will become overcrowded and the colony may abscond. These brood combs can be placed in another hive to strengthen its colony if necessary. One of the advantages of improved hives is the possibility on inspecting the colony to identify the problems and to take control measures. However, a precise time table for checking the hives cannot be given, because conditions vary.

9.3.3 Reading Frames or Top-bars

- Sealed brood should be compact if there are many open cells, it may mean some of the queen's eggs are not viable.
- Brood pattern: It should be in the center in a concentric semi-circle fashion.
- Ratios of egg : open larvae : capped pupae 1 : 2 : 4 ideal, which means the queen is good laying, the bees are sufficient in number to incubate those eggs.
- No eggs: If no eggs are found in the open cells, you can guess the queen has stopped laying or she is not present/and the colony is assumed to be queenless. In this case,

you should re queen such hive by giving: a queen cell or a new queen or join it in to a queen right hive (colony), uniting young larvae (<3 days old) & eggs from other strong and healthy hive.

- Supersedure: This indicates the queen is failing for some reason; leave the best two queen cells to hatch. Queen cups with lots of drones, sealed brood, no or small number of eggs are found the colony may swarm; take swarm control and prevention measures. If the hive is populous and crowded the queen cells are probably for swarm preparation. In this case, any of the following can be done: removing the queen cells, adding supers, splitting the colony, giving brood to weak colonies.
- Starvation: If there is no or the amount of honey and pollen is very small in quantity, it indicates starvation. Feed your bees.
- Combs: Replace combs that contain moldy pollen if there are more drone combs which are dried out and partly moldered away (decaying) which are broken or have large holes in them.
- Other observations: If the bees are volatile & change the behavior of the colony since your last visit it may indicate, lack of forage, pesticides use, pests, mites, disease, queenless, etc.

>>> SELF-CHECK QUESTIONS

Part 1. Multiple choice.

1. What are the materials required for fix foundation sheet?
   A. Wire   B. Frame   C. Embedder
   D. Hive tool   E. All, except D
2. One of the raw materials that is used for making comb.
   A. Honey   B. Pollen   C. Propolis   D. Wax
3. Which one of the following materials needed to make foundation sheet?
   A. Pure wax   B. Alcohol   C. Soap
   D. Wax moulding   E. All
4. Which one of the following is internal inspection?
   A. Look feed resource available   B. Egg laying
   C. Layout apiary site   D. Windbre
5. _____ is used to produce smoke, which causes the bees to consume honey, reducing their tendency to fly and sting.
   A. Bee veil   B. Smoker
   C. Personal protective equipment (PPE)   D. Glove
6. Which one of the following tools used for inspection of bee colony?
   A. Micro scope   B. Inspection hive   C. Smoker   D. Small box hive

7. Which is used to print foundation sheet artificially?
   A. Queen excluder
   B. Hive tool
   C. Casting mould
   D. Honey extractor
8. Which one of the materials is importance for baiting of hive?
   A. Cow dung
   B. Lemon grass
   C. Lime
   D. Wax
   E. All
9. All of the followings are importance use foundation sheet, except
   A. In frame hive beekeeping, foundation encourages the bees to build comb.
   B. Foundation enables bees to draw out comb more quickly.
   C. Reducing the effort and resources spent in wax and comb production
   D. Allowing more time and space for honey production
   E. None
10. It is an electrical device having 18-24 volts and is used to fix the foundation sheet to frame wires.
    A. Transformer
    B. Wax extractor
    C. Uncapping fork
    D. Hive tool

Part 2. Match the term in the left column with its definition in the right column.

1. Wax   A. Feed resources
2. Colony   B. Beekeeping equipment
3. Baiting materials   C. Queen and egg lay
4. Smoker   D. Bee product
5. External inspection   E. Social insects
6. Internal inspection   F. Cow dung

Part 3. Answer the following questions or statements in the space provided or on an additional sheet of paper if necessary.

1. What is the procedure make foundation sheet?
2. What is function of foundation sheet?
3. Write the two methods of inspection of hive.
4. What are the materials required to make smoker?
5. How do beekeepers extract wax?
6. Name the parts of smoker.

>>> REFERENCES

Kinati C, Tolemariam T, Debele K, 2011. Quality evaluation of honey produced in Gomma District of South Western Ethiopia [J]. Livestock Research for Rural Development, 23 (9).

Segeren P, 2004. Beekeeping in the tropics [M]. 5th ed. Digigrafi, Wageningen, the Netherlands.

MODULE 7:
MANAGING BEE COLONY

>>> INTRODUCTION

Managing bee colony is changing skill of beekeeper to make conducive environment for bee's colonies. Transition from novice to expert beekeeper is largely a matter of knowing how to nurture a struggling colony, learning to recognize and control the many problems that can arise and working with the bees so that they can realize their potential when conditions are good (Thomas, 2013).

Good management maintains strong colonies and allows for greater honey production and the potential to split and increase the total number of the beekeeper's colonies, all of which makes beekeeping much more profitable for hive owner.

To understand the relationship of climate to beekeeping, it is useful first to understand two concepts related to bees and their environment. These are the nectar flow and the honey flow. Although beekeepers often speak of these as being the same, they are different. This module covers of honeybee management practice like the regular inspection of colonies to assess the status of brood and worker bee condition, giving additional hive super, space reduction, feeding and maintaining colonies during dearth periods and detection and control of bee pests which enhances colony performance such as reduced absconding, improved colony strength and higher hive yields (Tolera and Dejene, 2014).

Therefore, the objective of this module is to provide basic information and technical procedures of colony management activities that critical for a beekeeper to increase his/her hive products.

1 PROVIDING SUPPLEMENTARY FEED

Honeybees store honey for their own consumption during dearth period. Beekeepers are harvesting honey, which the honeybees stored for themselves. As a result, honeybees face starvation due to lack of feed. To overcome this kind of problem, supplementary feed is required for the honeybees. The supplementary feeds that can be afforded for honey bees during starvation and manipulation for different operation in different areas of Ethiopian country are *beso*, *shiro*, sugar syrup, honey and water mixtures and peeper (Tessega, 2009). In addition to supplementary feeding, planting indigenous bee forage like *Tenadam*, *Girawa*, *Gesho*, sunflower, *Sesbania*, *Yekelem abat*, *Wolkef*, *Wanza*, *Eucalyptus*, and *Nuge* is also required to get the intended honey yield. Bee forage determines the amount of honey yield obtained. The existence of more bee forage results in high honey production provided that other factors are suitable for honey production. Plantation of multipurpose, drought tolerant, pollen and nectar rich plants and integrating these activities with the development of apiculture should be the major concern of the beekeepers at all levels.

The natural food of the honey bee consists of pollen, nectar and water. Proteins, vitamins, minerals, and fats are obtained from pollen. Pollen is essential to produce larval food and for brood rearing in honey bee colonies. Inadequate pollen stores in the immediate area of the winter cluster hinder brood rearing and the development of strong colonies. If colonies are found deficient in pollen stores early in the spring, you can extend their pollen supplies by feeding either pollen supplements or substitutes (Standifer et al., 1977). A pollen substitute is a protein source containing all the essential nutrients for bees but without pollen, whereas a pollen supplement is a protein source that has some pollen added to it. In early spring, before pollen and nectar are available or at other times of the year when these materials are not available for bees in the field, supplementary feeding may help the colony survive, or make it more populous. Feeding a hive need to build new comb, raise brood, and store food for those days they cannot gather nectar. Bees cannot live on pollen alone and there are times when the beekeeper must provide a substitute for nectar to prevent immediate starvation or to ensure sufficient stores normally for the winter and early spring but also at other times of the year and even in the summer (Food and environmental research, 2009).

This operation is known as 'feeding', which means giving the worker bee's access to a supply of suitable food placed in a suitable container. A suitable food for this purpose is sugar syrup or inverted sugar, and the container in which it is placed is known as a feeder. Food accepted by bees is either eaten or stored, if it is stored, it may be eaten subsequently or transferred to other cells.

The ability of the beekeeper to manipulate nutritional conditions in the hive by moving the hive to alternative floral sources or by supplementing or substituting protein or carbohydrates will make the difference between harvesting average honey crops and harvesting excellent honey crops (Doug, 2000).

1.1 Time of Feeding Honey Bees

We do not supplement honey bees with additional supplementary feed everyday like other livestock, but we provided them with supplementary feed in occasion when they are in need. Therefore, the honey bees are deliberately supplemented with recommended feed as the following occasions happened in and around the hive.

Bees should be fed in times of food shortage when there are no flowers. They should also be fed during drought or excessively wet, windy and cold periods when the bees cannot get outside. Feed bees when activity is low and in poor flying weather. If you see that traffic to and from the hive is slow, then the bees might need feeding. If the combs are dry and there is no honey, the bees are hungry. Feeding at such times may prevent the bees absconding and migrating away from the area or even prevent starvation.

Bees should be fed to replace the honey harvested from them at the end of the season, especially if a lot of honey has been harvested. Remember that a colony from which you have already removed a lot of honey cannot bridge a dearth period without being fed with sugar solution. A colony is fed to stimulate development during dearth periods and in preparation for the honey flow.

Regular feeding with very small quantities of sugar solution (or diluted honey) stimulates the development of brood. Bees should be fed for 6 – 8 weeks before a nectar flow when flowers are plentiful. The queen will then lay her eggs and the colony will build up in numbers before the honey flow. There will then be many bees ready to go out and collect nectar thus more honey to harvest.

Bees can be fed to assist them when establishing a new colony. Initial feeding when establishing a new colony will help a swarm or divided to settle down during their first nights. This will help to prevent them from wanting to abscond during their first few days.

Bees should also be fed at times of stress such as disease, sickness or after spray damage from insecticides. If the bees have been sick, remove any rotten, dry and dead larvae from the hive and then feed the bees.

1.2 Categories of Feeding Techniques

Stimulative feeding: Feeding of colony stimulates the queen to lay more eggs to ensure much brood rearing.

Manipulative feeding: This type feeding is used when some hive operation is carried out such as queen rearing, inspecting uniting etc. If the bees are fed either before or after the operation they are guided to the special manipulation.

Supplementary/emergency feeding: At certain time pollen shortage can retard colony development. During such time sugar syrup or honey would not materially improve the condition. The honeybees must have a supply of pollen or pollen substitute readily available to restore the depletion of food in the hive.

1.3 Feed Types for Feeding Honey Bees

The most important feeds for honeybees are honey, sugar syrup, pollen substitute, pollen supplements, dry sugar and homemade candy (Rogala and Syzmas, 2004).

1.3.1 Honey

Honey is the first choice for honeybees, and is free from disease spores. It is a natural and their own product. So, it is the principal food stuffs of honeybees. Feeding honey to a hive is not desirable if the aim is to stimulate the hive. The colony will reduce the brood area, the bees will become more defensive and robbing activity will increase. According to different scholars finding, feeding of honey may have risk in transferring disease associated honey especially American Foul Brood if the source of the honey is not known.

Procedure of feeding honeybees

1) Put the sugar solution into a bowl or jar and place some broken twigs/grass to float on the surface of the syrup to prevent the bees from drowning. The bees will sit on the grass to drink.

2) Use a specially designed feeder box (food grading bucket) for a top bar and frame hive.

3) Putting the feed in the hive

➤ Remove one or two top bars or frames from the hive and put the feeder in the place close to the cluster of bees so that they will find it quickly. Remove the feeder bottle as soon as it is empty. Place the feeder bottle close to the cluster of bees so that they will find it quickly. Beekeepers should make sure that there are no openings through which bees, wasps, ants etc. can steal the sugar and never offer food outside of the hive.

➤ Feed bees in the evening so the bees can get used to the presence of food during the night and will have settled down by morning. Never feed in the middle of the day. If you have several colonies, always take the food out of the hive during the day to prevent robbing and replace it at night. Alternatively feed all the colonies together during a one-time feed.

4) Stop feeding as soon as the bees no longer immediately take up the sugar or if it remains untouched for a day.

1.3.2 Sugar Syrup

This is prepared by dissolving ordinary table sugar (sucrose) in water. The ratio of sugar to water is depending on the objective or type of feeding. According to Yspuh and Solan (2012), sugar syrups is prepared and fed as follows:

➤ For general feeding, prepare syrup of one part of sugar and one part water (by volume), say one cup sugar and one cup water.

➤ For preparing stimulatory feeding, make dilute syrup by mixing one part sugar with two parts water.

➤ Feeding during autumn/winter should be with concentrated sugar syrup. To prepare heavy sugar feeding, dissolve two parts of sugar in one part of boiling water and add one tablespoon full of tartaric acid to 50kg of sugar, to prevent crystallization of sugar from the syrup.

Procedures of making sugar syrup

Only refined white granulated sugar should be used for making syrup. It is wasteful to feed syrup containing un-dissolved sugar, because the bees cannot take crystals that settle at the bottom of a feeder and if an inverted contact feeder is used, the crystals may clog the holes completely.

1) Take some sugar, some hot boiled water and a container.

2) Dissolve one part quality white crystal sugar and one part hot water together. Use boiled water to avoid disease but do not boil the mixture.

3) It is better not to use raw sugar or brown sugar as it may cause sickness among the bees.

4) Stir until the sugar is dissolved, as it is difficult for the bees to eat if there are whole crystals. Add a teaspoon full of honey if available.

5) Never prepare more sugar solution than the bees can take up in 2 days to minimize intrusion by robber bees and fermentation of the feed inside the hive.

Preparation of candy

1) Collect together the materials for preparing candy (honey, powdered sugar, water, container for mixing). Grind the sugar if powdered sugar is not available.

2) Mix 0.5kg powdered sugar and 100g honey thoroughly in a cooking bowl or bucket. You can use water if no honey is available, but it is not as nutritious.

3) Wrap the candy in wax paper or a plastic sheet keeping both ends open and place it on the top of the frame or on the bottom board in the hive, taking care not to squash any bees.

1.3.3 Pollen

It is the male germ plasma of plant. The honeybees obtain the whole of their requirement of amino acid, vitamins, lipids and minerals from pollen. It can be fed either alone or mixed with honey. It is for this season that beekeepers often trap pollen when it is abundant and feed them during periods of pollen shortage. Pollen is the major source of protein for honey bees. It is largely used to feed developing larvae and young bees to provide structural elements of muscles, glands and other tissues. It is also used in the production of royal jelly, which is a specialty food produced by worker bees that is fed to the queen, developing queen larvae, and worker larvae up to 72 hours of age.

A major factor which has been found to limit rapid increase in colony population is the insufficient supply of suitable pollen. A pollen with less than 20% crude protein cannot satisfy a colony's requirements for optimum production. A beekeeper who wants to harvest pollen must use 'a pollen trap' to remove pollen carried by the bees before they enter the hive; the colony continues to need pollen, and foragers therefore continue to collect.

1.3.4 Pollen Substitute

'Substitutes' suggests that either nectar or pollen, or both, are completely deficient in the field. Honey or nectar substitutes are usually in the form of sugar, preferably as sugar syrup. Large quantities of thick syrup are suitable for feeding to bees to store for winter, whereas small quantities of thin syrup fed regularly stimulates the colony to expand the brood area. If the purpose is to stimulate the colony and increase population numbers, then attention to the protein components of the diet is also essential.

Pollen substitute is any type that can be fed to bee colonies to replace its need for natural pollen. Among the most commonly used protein source from which pollen substitute can be prepared is soybean flour, brewer's yeast and dried skimmed milk. The recommended mixture is 3 : 1 : 1.

1.3.5 Pollen Supplement

The term 'supplements' suggests that there is already some naturally occurring pollen and/or nectar in the area for the bees and the beekeeper is making up the shortfall by feeding the hive strategic supplements. Supplements should contain the nutritional components that are deficient in the field as well as make up the required volume a colony may consume.

A pollen supplement is a pollen substitute that contains about 10% natural pollen (dry weight basis). Pollen supplement diets containing 20 percent or more of either soybean flour or brewer's yeast are highly palatable to bees and have the nutritive requirements for their growth and reproduction (Standifer et al., 1977). To prepare the pollen supplement diet as moist cakes for feeding inside the hive, first dissolve the pollen pellets in water since they do not readily soften in sugar syrup. Then, stir in the sugar until it dissolves or is well mixed with the pollen. Finally, add the soybean flour, wheat, or brewer's yeast to the water-pollen sugar mixture and stir thoroughly. The bulk supplement should be made up into 1.5 pounds cakes wrapped in waxed paper to prevent moisture loss.

2 DIVIDING AND UNITING COLONIES

You can make a division of an existing healthy colony to populate a new hive but always choose the most productive and docile colony. By dividing it, you are spreading its good genetic characteristics. Sometimes we may also need to unite colonies. Beekeepers unite colonies to enlarge a colony and improve their yield of honey or to survive the dearth (Segeren, 2004).

2.1 Dividing a Colony

Colony division is a method of multiplying bee colonies and it is the other method of queen rear in, i.e. producing two or more colonies from a mother colony. Colony division is used to control swarming, as well as in commercial beekeeping to increase the number of colonies. The colonies can be used to increase the number of colonies in the apiary for honey production or sold for income.

Choose the strong, productive and less defensive colony to make divisions to increase your colonies. You can make a division of an existing healthy colony to colonize a new hive. Make division after the honey flow to increase colony numbers. The best time to divide a colony is when the bees are getting ready to swarm. Avoid making divisions during the honey season because it will reduce the amount of honey to be produced. Between the beginning and the peak of the flowering seasons, strong colonies can suddenly become overcrowded with clusters of bees near the entrance, and large numbers of drones.

Dividing controls swarming and saves the beekeeper from losing the bees or the trouble of catching a swarm. Before dividing colony, the beekeepers should check the colony status. The mother colony selected for division should be strong and healthy. A strong colony means 10 frames covered with bees of which 6 contain brood, and sufficient stored food (honey and pollen). Then they must decide the season and time of dividing colony. Usually the best time for colony division is during the honey flow season. According to geographical location, colony division can be performed twice a year.

Materials and equipment

- All required personal equipment (bee veil, over all, boot)
- Empty hive box
- Water sprayer
- Hive tool/chisel
- Smoker
- A strong mother colony
- A frame fitted with comb foundation and empty comb
- Feeder/sugar
- Colony inspection equipment

Steps of dividing colony

1) Select the most appropriate mother colony.

2) Move the hive about 1 foot (30cm) to the left of the existing location.

3) Place an empty hive about 1 foot (30cm) to the right of the previous location, leaving the old location empty.

4) Take 3 to 4 brood combs from the mother colony together with the existing queen and place in the empty hive.

5) Keep 1 mature queen cell with 3 to 4 brood combs in the mother colony.

6) Divide the combs with food stores equally between the hives. Remove any remaining queen cells.

7) Divide the adult bees equally between the hives.

8) Check whether the incoming foragers are entering both hives equally.

9) If more foragers are entering one of the hives, move it further from the previous location and move the other hive closer to the previous location. Continue to adjust until equal numbers of foragers are entering both hives. Add frames with empty combs or comb foundation to the colony with the queen after colony division.

10) Close and cover the hives.

11) Divided colonies can be moved to the desired position by increasing the distance from the old position at a rate of 1 to 1.5 feet (30 to 45cm) per day in the evening after the bees have stopped foraging.

12) Divided colonies should be fed with sugar syrup in the evening for 3 days after division and comb foundation added as necessary.

2.2 Uniting Colonies

Beekeepers unite colonies to enlarge a colony and improve their yield of honey or surviving the dearth, control a worker-laying problem. A colony can produce surplus honey only if it is strong enough and contains 6-8 combs with plenty of brood and sealed honey and covered well by bees. This very much depends on the colony having a productive queen. If a colony fails to produce surplus honey for 2 seasons, or if it is weakened by repeated swarming, then it can be strengthened. Two weak colonies can be combined to make one strong colony. One large colony collects more honey than 2 smaller colonies. A colony can be united with another colony or with a swarm.

Before uniting colonies, however, it is essential that you know the reason why you are doing this. For example, if a colony is queenless or weak, you should know why. It would be pointless uniting this hive with another if it had a disease: you would be giving the disease to the healthy hive. So, before uniting colonies, check for disease. The problem with uniting colonies is that you are trying to combine two units of bees that will immediately fight each other when you put them together and, in the process, you will lose lots of bees and possibly one or both queens (Segeren, 2004). You must convince the bees they are not enemies so that they unite peacefully.

Reasons for uniting colonies

The colonies are recommended to unite if they have the following indication:

- Weaker colonies: to make a single strong colony
- Queen-less colony or weak queen: The queen-less colony should be united with a colony with a good queen (a 'queen-right colony').
- Worker laying: Sometimes worker bees may lay. The laying workers should be removed as soon as they start egg laying and the remainder of bees united with a queen-right colony.

- Inability of the queen to lay fertilized eggs
- To increase honey production: Two or more colonies can be united at the onset of the honey flow season to increase colony strength and maximize honey production.

Materials and equipment

- Personal protective equipment (bee veil, boot, over all)
- Newspaper
- Weak colonies
- Water sprayer
- Cell bar
- Smoker
- Chisel

Methods of uniting bee colony

You can unite bee colonies in two main ways:
- Earlier methods of uniting were based on the masking of the colony odour by sprinkling the bees with sugar syrup or even dusting them with powder. Sprinkling sugar syrup is still used and probably owes its success as much to the bees being occupied licking up the syrup as it does to the masking of hive odour.
- The most common method of uniting two stocks of bees is the so-called newspaper method. In this method the two colonies selected for uniting are joined by removing the lid of the one and the bottom board of the other, placing one colony on top of the other but separating them by a sheet of newspaper (Fig. 18).

It is very convenient to hold the newspaper down with a queen excluder, which also tends to retard the rate of mingling of the bees. It is usual to punch a few small holes in the paper to facilitate the merging of the two stocks. The bees chew through and remove the remaining paper as they mingle to form a coherent unit (Fig. 7.1).

3 MIGRATING COLONIES

In migratory beekeeping, the beekeeper transports the bee colonies to locations where flowering plants are present. This practice is encouraged in areas where the bee forage is spread over a large area of land. The migratory practice allows the beekeeper to harvest honey more than twice a year and thus to yield more honey. However, this practice requires that the beekeeper knows the time of flowering in different locations to plan the movements well during the year. Timely and safe movement of the colonies is important (FiBL, 2011).

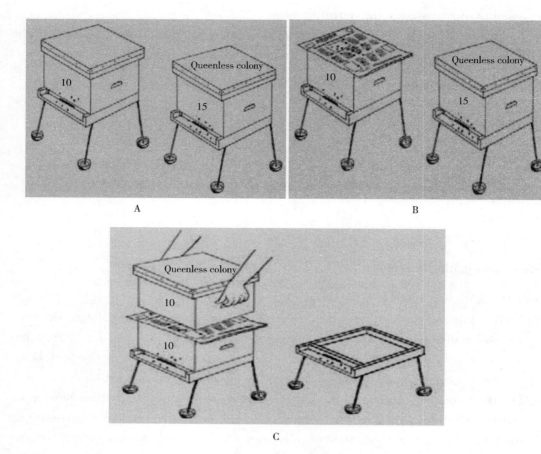

Fig. 7.1 Newspaper method
A. Queen-right and queen-less colonies moved close together
B. Covers of queen-right colony with newspaper
C. Queen-less colony placed on sheet of newspaper above queen-right colony

4 SUPERING AND REDUCING THE HIVE

4.1 Supering

Supering is defined as the additions of hive bodies on the existing hive which contain bee colony. Putting new frames alternatively with old frames in both base and super boxes. As soon as flowering commence, the queen start lay eggs up to 1,500 - 2,000 a day during flowering period. As a result, the population become high and the population may become congest. If supering is too late, the colony become swarm due to overcrowding which brings in reduction of population and decreases in honey yield as well as bees may spend much of their time in building the comb instead of collecting and storing honey.

If supering is too early, it may become difficult to the bees build the new combs and to manage the unnecessary space created to them. The newly added hive body is the place where bees congregate when working and helps to distribute the populous colony not to be confined in the brood area & prepare themselves for swarming.

If not supering when needed, often results in the substantial losses because bees become diverted from honey storage to swarm preparation.

Rule and time of supering

➢ As soon as the bees are fulfilling their brood area.
➢ As soon as the bees are working in the first super to the extent of about 3/4 of its capacity a second super should be given.
➢ Added just before (at the time) the tops of the frames are whitened with fresh bee wax.

Supering procedures

Beekeepers usually lay a queen excluder between the chambers to prevent brood rearing in the honey super. The disadvantage of this is that it limits the brood nest, which leads to extra swarming, less honey production and more absconding later. A hive like this is too small for large colonies to develop in. Therefore, it is advisable to work with more than two chambers.

Fig. 7.2 shows the supering procedures. The sequences are first put the first super on the brood chamber, where queen excluder is placed between them. The second super is placed between the previous (first) super and the brood chamber.

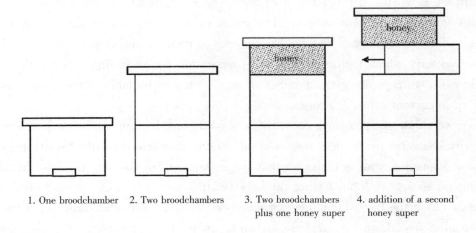

1. One broodchamber   2. Two broodchambers   3. Two broodchambers plus one honey super   4. addition of a second honey super

Fig. 7.2 Supering procedures

4.2 Reducing Super

It is a bee management practice that even under normal condition removing of boxes or a box from one set of colonies is a common work in bee husbandry. This is usually practices of colony management during the dearth period. Its management depends on the strength of the colony. Desupering (reducing super) is carried out when the honey crop has been removed and colony does not occupy the super. Because extra super harbors pests, it is a step taken to remove added supers one by one as the decrease of bee population occurs during the long dry months where food insecurity is high to bees. Reducing box until a base hive is left during the offset seasons is the appropriate practice in beekeeping.

Rules and time of reducing super

The reducing super is will be occur if the colony shows the following indication:
- Decreasing of population
- Having of unoccupied combs
- Complete or partial moving of bees either to the bottom or to super boxes.

During reducing supers, instead of removing all frames from the top, select and reduce only empty combs (without nectar, pollen), old combs, broken frames from both base and super boxes. After reducing if the combs are not too old and or damaged, hang them in aerated areas to be reused. If the combs are old and damaged, strain the combs to be used for foundation sheet making for re-super. Then clean the frames and the box to be re-super in the next active period.

5 CONTROLLING COLONY SWARMING

Swarming is a natural propagation of honey bees. A separate group of bees with one (or more) queens is called a swarm. The event of a swarm leaving the colony is called swarming. This is the normal way for bee colonies to increase and disperse in the area. It is not known exactly which factors trigger the preparation for swarming. Too little space in the brood nest, thus limiting the number of eggs that can be laid by the queen, is often given as an important cause (Segeren, 2004).

When we think about swarm control, it should consider the following two activities: First, there is swarm prevention and, second, there is swarm control. Swarm prevention is all about managing your colonies so that the swarming impulse doesn't arise in the first place. Swarm control is about letting the bees swarm, but only under your control so that you at least retain the bees. This may occur when you have left it too late to prevent them from swarming or, despite your best attempts at swarm prevention, the bees are still determined to swarm-and it happens.

The most frequently ways of controlling reproductive swarming by beekeeper is removal of queen cell, killing queen of the swarm and reuniting of honeybee colony to its mother, supering, and using large volume of hive as colony increase.

5.1 Contributing Factors for Swarming

Swarming is an instinctive desire of honey bees to increase their numbers by reproducing at the colony level, doubling their chances of survival. We don't fully understand this behavior, but we know of some contributing factors. These are as follows:

- ➤ *Seasonal changes*: Environmental can stimuli bees for swarming. Rising average temperatures and generally warmer weather after the winter period will stimulate a colony to swarm in spring.
- ➤ *Congestion*: Congestion is said to reduce the amount of queen pheromone available to an individual worker bee, thus giving it the stimulus to begin constructing queen cells. Congestion occurs due to the prevailing breeding situation and the restricted space provided for the colony. Managed bee hives often are restricted artificially to a single brood box—this is said to lead to a higher incidence of swarming (Doug, 1999). Preparation for swarming starts with the building of swarm cups. These are short bowl-shaped cells which have their openings facing downwards. These swarm cups are usually found at the bottom edge, but also at the front and back edges of the comb.
- ➤ *Old queens*: Although not the only cause of swarming, old queens are more likely to swarm.
- ➤ *Genetics*: The genetic variation between strains and races of honey bees has significant effect on swarming. Some colonies under identical conditions will be more inclined to swarm than others.

To check if a colony is getting ready to swarm and wanting to divide itself, we must look for signs that the colony is overcrowded, and the queen has run out of cells to lay eggs in (Adjare, 1990). A colony can fill between 9 and 15 brood combs with brood of all stages, including a lot of drone brood and sometimes there is even little surplus honey. There will be clusters of bees outside the hive and lots of drones flying. Also, the bees will be producing queen cells (the long thumb shaped cells protruding from the edge of the combs). Dividing stops them from swarming and saves the beekeeper from losing the bees or the trouble of catching.

5.2 Preventing Swarming of Honey Bees

Swarm prevention should be initiated before the first significant nectar flow begins. Recognizing the signs of the approaching swarm season is crucial to preventing a swarm from happening. There are numerous practices a beekeeper can employ to limit or prevent swarming. Making sure that the queen has enough room to lay eggs. Diligent monitoring of the hive during early spring through the nectar flow is the most important part of each of the following described practices (Sara et al., 2012). Make extra space around the brood nest by removing honeycombs and putting in empty combs near the brood nest (Harlan,

2001). Weakening the strong colony can prevent its urge to swarm. Destroy all the queen cells in the colony then switch the hive location with a weaker colony. Artificially swarming the bees for swarm control by making a division. Making divisions is also a great way to increase your colonies. Generally, adding super early/on right time, inspecting colony and damage queen cells, inserting queen excluder to partially limit the egg laying of a queen, and removing brood & giving to weak colony are the ways how you can prevent swarming.

Methods to prevent swarming

Good swarm-prevention methods should reduce swarming with a low degree of colony interference and should be compatible with good colony management for both pleasure and profit. When considering your swarm-control strategy, try to think in terms of employing the following manipulations in conjunction with each other, not as isolated examples.

1) Re-queening annually (or at least every two years)

This is one of the best methods for limiting swarming in your colonies, especially if you are a commercial beekeeper and have perhaps thousands of hives. It is difficult under these circumstances to keep such a close eye on matters but, if you re-queen annually, you will at least know that even in your absence the number of colonies swarming in your bee yards will be low. For a beekeeper with only a few colonies it is an easy method to employ. Research has shown that a queen under a year old with plenty of queen pheromone is much less likely to swarm than a queen in her second year; a two-year-old queen is less likely to swarm than a three-year-old; and so on. The figures are quite remarkable. Re-queen, therefore, *every* year so that no queen is over 12 months old. Fall or autumn re-queening is perhaps more difficult than spring re-queening, but it has so many advantages over spring re-queening. From my own experience I think the reasons for this are the better weather and the larger numbers of mature drones around. Troublesome spring weather and the chances of fewer drones being available make spring mating less certain. The great majority of professional honey producers re-queen every 12 months. They do this for a reason: less swarming, more eggs/bees.

2) Supering up

This involves putting honey supers on to the brood body(s) in time for the honey flow. The first box should be filled with comb, especially if the season is early—bees have difficulty producing wax early in the year. Putting supers on in time is not only essential for honey storage preparation but it also limits swarming by giving the bees more room in the hive.

6 PREVENTIONS OF ABSCONDING

Absconding is the abandoning of a hive by a colony. Usually absconding occurs either by seasonal conditions inside the hive or by disturbance, plus a combination of the two. Mismanagement of colonies in the later season cause natural absconding. Honeybee colonies abscond when they fall under a critical weight of decreasing nectar flow, or if conditions are not ideal due to lack of forage or water, or excess sun or wind. To prevent absconding harvesting should be such that the colony will not fall later below this critical weight. One or two full honey combs should be left at the side of the brood nest. Honey production and number of colonies, i. e. occupation of hives, will increase.

Absconding is common in tropical species and races of honeybee. Leaving some honey for the colony at harvesting can reduce absconding. To prevent absconding the beekeeper should make the environment ideal for bees in usual activities (FiBL, 2011). Regular inspection of the status of the colony and investigating any stress factor help to protect honey bees from unnecessary absconding. Understanding the causes and reduction of absconding leads to increased production; modern honeybee husbandry is meant to have a higher production than from the wild. The best way to avoid absconding is to protect the bees from disturbances and to ensure a certain amount of food (at least four or five combs full of honey) is provided (Segeren, 2004).

7 INSERTING AND REMOVING QUEEN EXCLUDER

A queen excluder is a device used above the brood chamber through which worker bees can pass, but queens and drones excluded because of their size. Queen excluders are made of various metals or plastic. The most common use of a queen excluder is to place it directly above your top brood box. Then, all honey supers are placed above the excluder. The excluder then can keep the queen in the brood chambers and excludes her from getting into the honey super and laying eggs. it is inserted during supering hive.

The proper way to place a queen excluder on a hive is with the support wires on the bottom side of the queen excluder. Queen excluders are removed in the autumn; otherwise, the queen would not be able to move with the winter cluster and would die from exposure. The death of the queen in winter would doom the hive unless the beekeeper introduces a new queen in the spring. Queen excluding must be inserted 21 - 30 days before the expected honey-harvesting period.

When inserting queen excluder, force the bees including the queen to the base box; take out each frame one by one while brushing or shaking away the bees to the box. When the super is free from combs and bees, remove the super and check frames with honey & nectar or sealed brood from the base and replace with combs with young brood or pollen from the top box. Then put the queen excluder on top of the base hive & place the super and put combs with honey, nectar & sealed brood above queen excluder.

8 TRANSFERRING BEE COLONY

To undertake commercial beekeeping and maximize profit, colonies must be transferred from traditional to modern hives to enable improved colony management practices to be applied. It is difficult to increase income from traditional beekeeping in log or wall hives because the amount of honey produced by bee colonies in traditional hives is very low and proper colony management is not possible. It is possible to transfer bees from a wild nest or from a traditional hive with fixed combs (combs which are not moveable) to top bar hive and improved hives. Transferring is not easy for the beginner. Transferring bees from a fixed comb hive is much easier than transferring bees from a wild nest.

8.1 Season (Time) of Transferring

The best season to transfer colonies of bees from one hive to the other is a period of honey flow. An ideal time is during honey bees plant bloom at the first week of active season (at the end of heavy rainy season or the end of dry season) in order that the bees could exploit the environmentally available flowering plants and build up their population.

The best time to do this transfer is shortly before sunset, not in the middle of the day. In the middle of the day, many worker bees would have been out of the hive foraging but in the evening, they will have come back.

Tools, material and equipment required

- Strong colony
- Empty hive either transitional or improved
- Smoker
- Bee brush
- Water sprayer
- Mat/iron sheet/sack
- Knife
- Queen catcher
- Petri dish/container
- PPE
- Feeder and feeds

Procedures transferring bee colony to modern hive

All procedures and preconditions for colony transfer should be followed as described. Prepare modern hive with movable frames to transfer colony from the traditional one. A hive tool, a smoker, strong cord, a sharp knife, etc. are the tools needed to make the transfer. The colony, which is going to be transferred, should be placed in its new site (place) at least three days in advance. This is very important to orient the bees to their new environment and foraging direction. The site (place where transferring will take place, and colony will be placed) should be cleaned. The smoker should be filled with fire and smoking materials at least 10 minutes in advance. The beekeeper must prepare his/her protective clothes dress them properly. Move the old hive/traditional hive to cleaned and prepared site and place it on the ground where the mats/plain iron sheet, or canvas spread in horizontal position. Place one of the hive, bodies with frames of comb and foundation face to face with fixed hive/traditional which should fit together without opening or cracks. Smoke the old/traditional hive and tap on the sides with hands. Continue smoking and tapping (drumming) to drive the bees up into the new hive body. Drive all the bees up into the new hive body, spray water on bees to prevent flying as possible by colleagues. Remove the comb from the old hive, carefully cutting the comb from the sides with a sharp knife and handle the comb carefully to preserve the brood. Watch and inspect for the queen as you drive and remove the combs carefully. If you get queen, take off your hand glove and catch her with your fore finger and thumbs at her thorax, then put her in queen catcher and place in new hive. As you remove combs from the old hive, shake down the bees on mats. Finally, make sure all bees are transferred to new hive and settled, then cover the new hive and safely turn back where it isolated previous for transferring and place for production.

8.2 Follow up after Transferring

Once the colony is transferred to the new hive the beekeeper should undertake external hive inspection regularly whether the transferred bees are fully established. This can be indicated by strong movement of bees around entrance and being busy with their usual works. Inspect the colony internally to check whether the colony started comb building. While internal inspection if you observe the comb built newly and worker bees are being busy with building combs, it is recommended to release queen from queen cage to colony to start laying eggs. On the next consecutive days, observe or examine for the egg laid by the queen. This convenient indication for queen/colony being settled down after transferred. Then carry out the necessary hive manipulation for the building up of the colony fortnightly once in a month.

>>> SELF-CHECK QUESTIONS

Part 1. Choose the correct answer from the give alternatives.
1. Which of the following is/are the reason for uniting of bee hives?
    A. Weak colonies   B. Queen-less colony
    C. Failing queen   D. All
2. One of following alternative is not stimulating factor for swarming of colony

A. Overcrowding
B. Poor ventilation
C. Having low space for egg laying
D. Damage newly emerged queen cells

3. A tool which placed between honey chamber and brood chamber to exclude down the queen and drone is
   A. Supper
   B. Queen cage
   C. Queen excluder
   D. Queen catcher

4. _____ is type of feeding when operation is carried out.
   A. Emergency feeding
   B. Supplementary feeding
   C. Simulative feeding
   D. Manipulative feeding

Part 2. Fill the blank space.

1. _____ is the process of adding hive box with full components
2. _____ is recommended season and _____ the best time for uniting of bee colony.

Part 3. Essay part.

1. Write the occasions when bees need supplementation.
2. What causes bee colony prone to swarm?
3. How can you increase the production of honey per hive?

>>> REFERENCES

Cramp D, 2008. A practical manual of bee keeping [M]. Spring Hill House, United Kingdom.

Gebremichael B, Gebremedhin B, 2014. Adoption of improved box hive technology: Analysis of smallholder farmers in Northern Ethiopia [J]. International Journal of agriculture and extension, 2 (2): 77 - 82.

Kumsa T, Takele D, 2014. Assessment of the effect of seasonal honeybee management on honey production of Ethiopian honeybee (*Apis mellifera*) in modern beekeeping in Jimma Zone [J]. Research Journal of Agriculture and Environmental Management, 3 (5): 246 - 254.

Rogala R, Syzmas B, 2004. Nutritional Value for Bees of Pollen Substitute Enriched with Syntheticamino acid [J]. Journal of Apicultural Science, 48 (1).

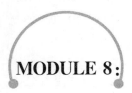

MODULE 8:

HARVEST, PROCESS AND STORE BEE PRODUCTS

>>> INTRODUCTION

Primarily, we keep bees to produce honey and other products, such as propolis, bee venom, beeswax, pollen (bee bread), royal jelly (bee milk) and brood. The bees collect nectar (plant sugar) from flowering plants and store it in their combs by reducing its moisture content to the level they want for their future consumption. This stored form of nectar is called as honey. Thus, honey harvesting is collecting of the stored reserve food for bees and their young (brood) from their nest. Six bee products are known. These are honey, beeswax, propolis, pollen, royal jelly, and bee venom. This chapter enables to identify and characterize honeybee products, and to harvest, process, package and store honeybee products.

1 IDENTIFY AND CHARACTERIZE BEE PRODUCTS

Apart from the importance of honeybees as the basis of agriculture in the pollination of seed, fruit, vegetable and legume crops, they are also of immense importance to the beekeepers in production of honey, beeswax, pollen, royal jelly, propolis and bee venom. Whatever found in the hive, which have direct or indirect contact with bees are called hive products. Natural products carried into the hive by honeybees and subsequently processed within the hive are: nectar, honeydew, pollen, propolis and water. Some of them have potential economic value to farmers and others. Another hive products, which are manufactured by bees within the hive, are: beeswax, royal jelly, bee venom and brood (bees). All of them have potential value in bringing economic return to the owners and they have very essential role in human medicine, human diet and in economy building.

The physical, chemical and uses of the honeybee products are described in the subsequent sections.

1.1 Honey

Honey is the natural sweet substance, produced by honeybees from the nectar of plants or from secretions of living parts of plants, or excretions of plant-sucking insects on the living parts of plants, which the bees collect, transform by combining with specific substances of their own, deposit, dehydrate, store and leave in honeycombs to ripen and mature (Codex Alimentarius, 2001).

1.1.1 Chemical Composition and Physical Properties of Honey

Chemical properties of honey

Honey consists mostly (80%) of sugars that are readily absorbed by the body (FAO, 1990), moisture range is 12.2 – 22.9%; above 18.6% it is disqualified, but it is safe from fermentation below 17.1%. Below 15% it is graded low score because it is too thick. Bees add enzymes to break sucrose to simpler sugars that aids digestibility. Glucose and fructose predominate and give it most of its sweetness, energy value, and physical characteristics (White, 1975).

Acidity of honey masked by sweetness, but adds to its complex flavor, its pH value is about 3.91. Minerals are about 0.17% weight; darker honeys tend to have more minerals. Honey contains some enzymes: invertase added by bee converts nectar sucrose to glucose and fructose; another is glucose oxidase added by bees which helps keep nectar from spoiling while it is ripening.

Protein and vitamins are only at trace levels. Granulation may increase as ratio of glucose to water increases, the honey can start to ferment during storage if the water content is too high (>19%). Fermentation can be prevented by heating the honey to a temperature of 55 or 60℃ over a period of 8 hours, followed by rapid cooling. However, heating the honey for too long will cause both the taste and smell of the honey to deteriorate. Heated honey is of an inferior quality as the enzymes are broken down. A more detailed definition of the composition of honey is shown in Table 8.1.

Table 8.1 Chemical composition of honey

Composition	Percentage (%)	Mean (%)
Water	12.2 – 22	17.0
All sugars	78 – 88	80
Fructose	21.7 – 53.9	38.2
Glucose	20.4 – 44.4	32.9
Sucrose	0.0 – 7.6	2.3
Maltose	0.1 – 16.0	7.5
Acids (succinic, gluconic, acetic citric and butyric acids)	0.2 – 1.0	0.57
Protein	0.1 – 0.4	0.26
Minerals (K, Na, Ca, Hg, Fe, Cu, Mn, Cl, P and S)	0.17	0.17
Pigments, flavor, aroma substances, and sugar alcohols	2.2	2.2
Enzymes (invertase, diastase, Glucose oxidase and acidic phosphate)	Trace	

Physical properties of honey

1) Granulation

Dextrose, a major sugar in honey, can spontaneously crystallize from any honeys in the form of its monohydrate. This sometimes occurs when the moisture level in honey allowed to drop below a certain level. Since the granulated state is natural for most of honey produced, processing is required to keep it liquid. Careful application of heat to dissolve 'seed', crystals and avoidance of subsequent 'seeding' will usually suffice to keep a honey liquid for six months. Damage to color and flavor can result from excessive or improperly applied heat. Honey that has granulated can be returned to liquid by careful heating. Heat should be applied indirectly by hot water or air, not by direct flame or high-temperature electrical heat, stirring accelerates the dissolution of crystals. For small containers, temperatures of 140 °F for 30 minutes usually will suffice.

If unheated honey can granulate naturally, several difficulties may arise. The texture may be fine and smooth or granular and objectionable to the consumer. Furthermore, a granulated honey becomes more susceptible to spoilage by fermentation, caused by natural yeast found in all honeys and apiaries. Quality damage from poor texture and fermented flavors usually is far greater than any caused by the heat needed to eliminate these problems.

Deterioration of quality

Fermentation of honey is caused by the action of sugar-tolerant yeasts upon the sugars dextrose and laevulose, resulting in the formation of ethyl alcohol and carbon dioxide. The alcohol in the presence of oxygen then may be broken down into acetic acid and water. As a result, honey that has fermented may taste sour. The yeasts responsible for fermentation occur naturally in honey, in that they can germinate and grow at much higher sugar concentrations than other yeasts, and, therefore, are called 'osmophilic'. Even so there are upper limits of sugar concentration beyond which these yeasts will not grow. Thus, the water content of a honey is one of the factors concerned in spoilage by fermentation. The others are extent of contamination by yeast spores (yeast count) and temperature of storage.

Honey with less than 17.1 percent water will not ferment in a year, irrespective of the yeast count. Between 17.1 and 18 percent moisture, honey with 1,000 yeast spores or less per gram will be safe for a year. When the moisture is between 18.1 and 19 percent, not more than 10 yeast spores per gram can be present for safe storage. Above 19 percent water, honey can be expected to ferment even with only one spore per gram of honey, a level so low as to be very rare. When honey granulates, the resulting increased moisture content of the liquid part is favorable for fermentation. Honey with

> high moisture content will not ferment below 50 °F or above about 80 °F. Storing at temperatures over 80 °F to avoid fermentation is not practical, as it will damage honey.

2) Sweetness

Honey sweetness depends on high fructose content and acidity. A few plants give bitter honey: *Agave* sp. (sisal), *Datura* sp., *Euphorbia* sp., *Senecio* sp. – isomer societies (for example, in East Africa), these honeys are very popular.

3) Hygroscopicity

It the ability of honey to absorb or remove moisture from the air. Honey, especially when rich in fructose, is very hygroscopic when the container is not closed. This may lead to an increase in water content and possible fermentation. For this reason, it is important that honey is always stored in containers with tight fitting lids.

4) Color, smell and taste of honey

We call the color, smell, taste and viscosity of honey its organoleptic or sensory characteristics. The taste and smell of honey are primarily determined by the flowers and plants the honey is made from. But these characteristics can be influenced by changes that take place in the comb, especially in combs that once held brood if honey is stored in them for a long time. For the consumer of honey, the important features of honey are its aroma, flavor, color and consistency, all of which depend upon the species of plants being visited by the bees.

For example, bees foraging on sunflower will produce a golden honey that granulates (crystallizes) quite quickly, while bees foraging on avocado produce a dark honey that remains liquid over a long period. The factors of aroma and flavour of honey are subjective, and honey is often judged according to its color. Usually dark-colored honeys have a strong flavour while pale honeys have a more delicate flavour. A great number of different substances (alcohols, aldehydes, organic acids, and esters) contribute to the flavour of honey. These are volatile compounds and evaporate easily at temperatures above 35℃: This is one of the reasons why honey quality is reduced by heat.

It is impossible to give a comparable value to the subjective values of flavour and aroma: The relative popularity of dark and light-colored honey varies from country to country. Color can sometimes be a useful indicator of quality because honey becomes darker during storage, and heating will darken honey. However, many perfectly fresh, unheated and uncontaminated honeys can be very dark.

5) Viscosity

It is a resistance to follow and affected by moisture/water content, the quantity of non-sugar and the temperature. The more moisture contents the less viscosity of honey and the warm the temperature the less viscosity.

1.1.2 Uses of Honey
- In nutrition: Honey is high energy carbohydrate quickly and easily digested.
- In commercial: keep the shelf life of packed goods
- In alcoholic drinks
- In cosmetics for skin, hair, face, hand, and body cream preparation: For preventing and curing rough and chopped hands and lips, and preserves hair color unchanged making lotion and shampoos cream etc.

Honey for medical purpose for dressing of wounds, serve as good appetizer and treatment for FMD.

1.1.3 Honey Quality and Categories

Honey quality

It does not matter where they are living in their own nest built in the wild or in any type of hive —bees always store clean and perfect honey. The place where they live has no effect upon the quality of honey that bees make. It is only subsequent handling by humans that leads to reduction in quality; if the honey is harvested when the water content is still too high (honey is still 'unripe'), if it is contaminated, over-heated, over-filtered or spoiled in any other way.

Honey categories

1) Honey categories concerning origin

Honey may be categorized according to its origin, the way it has been harvested and processed, and its intended use.

- *Blossom honey* is obtained predominantly from the nectar of flowers (as opposed to honeydew honey).
- *Honeydew honey* is produced by bees after they collect 'honeydew' -secretions of insects belonging to the genus *Rhynchota*, which pierce plant cells, ingest plant sap and then secrete it again. Honeydew honey colour varies from very light brown or greenish to almost black, and is an important type of honey for producers in coniferous forest areas of Central and Eastern Europe.
- *Mono floral honey* is where the bees have been foraging predominantly on one type of plant, and is named according to that plant. Common mono floral honey types are clover, Acacia, lime (linden) and sunflower honey. Mono floral honey is priced more highly than poly floral honey. Light, mono floral honeys like orange blossom or Acacia, because they look so attractive, always obtain higher prices then blends of honeys.
- *Multi floral honey* (also known as poly floral) has several botanical sources, none of which is predominant, for example, meadow blossom honey and forest honey.

2) Honey categories concerning processing

Comb honey is pieces of honeycomb, as produced by the bees, where the beekeeper has done no processing to separate the honey from the beeswax. The beeswax comb, as well as the honey, is edible. Comb honey always fetches a very good price, as the consumer can be sure that the honey has not been contaminated in any way.

- *Strained honey* is honey obtained by straining honeycombs, to separate the honey from the bees wax.
- *Chunk honey* is a jar of liquid honey inside which is placed a piece of comb honey. This can look very attractive. It is important that the liquid honey is a type that is very light and clear, and will not granulate over a long period. Honeys from Acacia (*Robinia pseudo acacia*) are often used for this. This type of product depends on the right type of honeys and excellent packaging, and can achieve a very good price.
- *Extracted honey* is honey obtained by centrifuging honeycombs.
- *Pressed honey* is extracted by pressing honeycombs with or without the application of moderate heat.
- *Crystallized or granulated honey* is strained honey that has crystallized.
- *Creamed honey* is strained honey that has been seeded to start crystallization and then stirred to produce a honey of uniform, soft consistency. On an industrial scale, honey is creamed by the 'Dyce method' (Dyce, 1975). About 20 percent of fine crystallized honey is mixed with liquid honey and the crystals are allowed to grow at 14℃. This procedure stabilizes the honey consistency, and does not affect the honey's authenticity, as no foreign matter has been added or removed.

1.2 Bees Wax

This is an inflammable substance acquired after processing some honey combs. This wax is used as a solvent for some injectables. Beekeepers with the knowledge use it for making wax foundations. The bee consumes between 8 and 15kg of honey to produce 1kg of beeswax. Bees wax is a valuable commodity that can be used to make a variety of products. It should not be thrown away (Fig. 19).

1.2.1 Wax Properties and Composition

Beeswax is a natural product that consists of many insoluble fractions. This gives wax a melting trajectory rather than a melting point. The melting trajectory lies between 62 and 65℃ and a relatively high level of energy is required to melt it. The melting trajectory offers many advantages. It makes the wax pliable: It already becomes soft at 35℃. Beeswax can be extracted from the comb using the heat from the sun, steam or hot water. Beeswax is chemically inert. It can therefore be used to protect materials from chemical substances and from honey by covering them with a thin layer of beeswax.

Wax is also suitable for uses in which the active ingredient must be released slowly. Beeswax does not dissolve in water; this makes it suitable for waterproofing materials and cloths and for resist techniques. Beeswax does dissolve in organic solvents such as benzene, ether or chloroform, as well as in fats and oils through heating. The color of beeswax is determined by the pollen that the bees collect during the building process. New wax is usually white, but it can also be yellow to reddish-orange. With use, the combs become darker, even brownish-black after they have contained brood. Beeswax bleaches in the sun.

The wax produced by different species of *Apis mellifera* has the same composition, but some components are in a different proportion.

Bees are stimulated to produce wax when the queen needs space to lay more eggs when the colony is expanding during the buildup. Bees are stimulated to produce wax when there is surplus honey to be stored during the honey flow and a lack of honeycomb in which to store it. Beekeeping using traditional hives or movable combs can result in high yield of beeswax (National bee keeping training and extension manual, 2012).

1.2.2 Uses of Wax

Beeswax is a valuable commodity that can be used to make a variety of products. It should not be thrown away. Price of beeswax currently 3 times more than honey, naturally, products of clear colors are traded for higher prices. Bees need wax as construction material for their combs. They produce it in their wax glands, which are fully developed workers wax. Wax can be used for more purposes such as cosmetics, pharmaceuticals, hard body cream, soft body cream, candle, shoe polish and to make foundation sheet. Bees wax has a wide variety of uses. The most important use of beeswax is in beekeeping itself, namely for the production of artificial combs. Artificial comb foundation is made of moulded or pressed wax sheets with cells imprinted on them that the bees very quickly and economically (using very little honey) build into comb. A surplus of beeswax can be found mainly in countries where artificial comb foundation is not used. The cosmetic industry uses beeswax as an emulsifier and binding agent in oils and fats because of the high amount of energy required to melt it and its melting trajectory. This makes these cosmetics hard when cold, and prevents them from melting too quickly in the sun like solid fats. Moreover, they react perfectly to human skin. Beeswax is therefore frequently added to creams, salves and lotions. Lipstick and mascara, which normally contain more than 30% beeswax, are both quantitatively and qualitatively important uses.

1.3 Propolis

It is a sticky dark brown or black glue-like substance used by bees as a sealing agent to close crevices or holes on the hive. This propolis is waterproof and can be used by builders in place of bitumen putty on asbestos roofs. In the hive, propolis is an anti-bacteria agent. The propolis can be used in making remedies for diseases such as asthma, skin diseases, arthritis to name but a few. Bees make propolis by mixing glue from trees and other substances extracted from flower buds. The bees use to fill cracks in their hive, to seal the entrance hole when it is too large, to make the hive watertight, to glue the top bars to the hive body, to strengthen the thin borders of their comb and as an embalming material to cover any dead hive intruder which they cannot remove from the hive. This hive product has several pharmacological properties. For instance, it is used in preparations to treat some skin diseases, and research on other uses is going forward. It is also marketable abroad. Propolis is collected from plants and is very different from beeswax. It has excellent antiseptic properties and is valuable both internally and externally. Collect propolis carefully to avoid getting it dirty. It should not be crushed into a ball. Store in a clean, dry container and use either dry or as a tincture (Pam Gregory, 2012).

1.3.1 Properties and Composition

Propolis has its own specific properties: It is sticky, brown and fragrant. The bees use it to fill undesired holes or cracks in the walls of the hive and they polish their cells as protection for the future brood. Bees also use propolis to adjust the size of the opening into the hive. In a severe winter, they will make it smaller. They also smear it on the inside of their hive and use it to stick loose parts of the hive together. This can be an advantage if the hive is moved. They use it to embalm undesired invaders, such as dead mice. Bees also mix a small amount of propolis with the wax used to cap the brood cells. Propolis is not water soluble and does not allow air to get through. It is hard at low temperatures but flows out at temperatures above 35℃. The color of propolis can vary from dark brown to reddish or yellow.

1.3.2 Uses of Propolis

Propolis is used for healing wounds, as a 'natural antibiotic' taken in addition to antibiotics and to strengthen one's health and immune system. For external use, propolis is processed in nose drops, cough syrup, toothpaste, lotions, salves, creams, skin oils, shampoo and skin soap. Health care products that contain propolis are used for wounds, scars, infections, muscle ailments, eczema, psoriasis, warts, moulds and nail cuticles (fungi). For internal use, propolis powder is often mixed with honey. To make tablets and capsules, propolis must first be purified because the botanical waxes and beeswax normally present make it difficult to absorb the propolis in the digestive system. For homeopathic uses, raw propolis is extracted with alcohol, or ethanol, to make the so-called mother tincture. This is processed in nutrient supplements and health care products or diluted further for use as. Tincture does not dissolve in water, so the best way to take it is to drip it onto a crust of bread, a sugar cube or tablet. Drops of light tincture (10%) can be added to a glass of water. Chewing gum, capsules, tablets, cough syrup, and mouthwash are also available (Fig. 20).

1.4 Pollen

Pollen can be very useful to us. Bees are very helpful pollination agents. Plants whose pollination is facilitated by bees produce quality fruits and seeds. If collected from the cells on the combs, the pollen is very nutritious food that contains minerals, vitamins and

MODULE 8
HARVEST, PROCESS AND STORE BEE PRODUCTS

carbohydrates. Some suffering from indigestion can take pollen as a laxative.

The male reproductive agent of flowering plants is collected by bees and stored in comb cells. It is fed to the brood in the larval stage. Pollen is collected from beehives using pollen traps. These remove the pollen pellets from the corbicula (pollen baskets) on the hind legs of the foraging bee.

Beekeepers can collect pollen from hives and save it to feed to the bees at times when no plants producing pollen are in flower for the bees to collect and eat directly. In the developed countries, pollen is also used in some expensive dietary supplements, since it is believed to have valuable medicinal properties.

1.4.1 Properties and Composition

Pollen grains have a tough outer wall: the exine. This sometimes has barbs that allow it to stick well to the bee's hairs. This outer wall is covered in a layer of wax, which makes the pollen very difficult to digest and is also the reason pollen can become fossilized and remains intact in the soil for millions of years. Despite this hard-outer wall, bees make it slowly more digestible and eventually after several weeks make bee milk or royal jelly out of it for the young larvae. Each pollen load comes from one plant species.

The amino acid pattern of the proteins in pollen determines its biological value for the bees. Bees in a colony visit various plant species, so the multi colored mixture of pollen loads usually has a good composition if it is not dominated by a deficient type, such as the pollen of corn. When the forager bees return to the hive, the beekeeper can usually recognize the origin of the pollen by the color of the loads. The composition and health value of the pollen vary per plant species. By looking at the pollen under a microscope, it is possible to identify its plant family, genus and species and this is called *melissopalynology*.

Pollen contains lipids, essential oils, vitamin E (tocopherol), carbohydrates, peptides, short proteins or oligopeptides, amino acids, pantothenic acid, anthocyanins, carotenoids, flavonoids, ferulic acids and enzymes as well as many minerals such as iron, manganese, zinc and spore elements.

1.4.2 Uses of Pollen

- For plant breeding programme
- Fruit pollination
- For feeding bees (supplementary feed)
- For human diet and for domestic animals
- In cosmetics health care
- Treatment for hay fever
- Corrects loss of weight
- Enhances digestion and stomach up sets
- Skin disease and radiation are cured
- Improves breathing and sexual potency
- Helps in blood pressure to drop

1.5 Royal Jelly or Bee Milk

Royal jelly is also called bee milk; it is used by the bees to feed the queen bee and the young larvae less than three days old. It is secreted from the glands of the 5 – to 15 – day-old worker bee. Studies show royal jelly to be a good source of vitamin B. Like pollen, it is thought to have medicinal value and is therefore used in certain expensive preparations. Royal jelly is used in making medicine to treat various ailments. People who have suffered from prolonged illness can take this for fast recovery. However, it calls for great knowledge for beekeepers to obtain the jelly from the cells without forcing the bees to abscond the hive. Royal jelly has all the constituents of a balanced diet.

1.5.1 Properties and Composition

The bee milk for the queen is the most nutrient rich and is therefore called royal jelly. The queen also gets much more than the workers. This is partly why the queen becomes much bigger and stronger than the workers. She can live for a few years, and thus much longer than the 4 weeks to 6 months, depending on the season, that the worker bees live. The composition of bee milk depends partly on the bee bread and thus the pollen. It is rich in vitamin B_1, vitamin B_2, vitamin B_6, folic acid, inositol, pantothenic acid, vitamin C and vitamin E (tocopherol). Royal jelly also contains peptides, lipids, sterols, aromatic oils, carbohydrates, enzymes, anthocyanins, carotenoids, flavonoids, ferulic acids, as well as minerals and spore elements from the bee bread.

The acidic fraction royalisin makes royal jelly effective in combating a broad spectrum of bacteria, but not fungi. Royalisin contains gamma globulins, which are important amino acids in the immune system. This fraction also contains 16% asparagine, which is needed for tissue growth. About half of the fat fraction is made up of 10 – hydroxy-2 – decanoic acid (10 – HDA), which plays a role in growth, the hormonal system and the immune system. Fresh royal jelly contains 2 – 15% 10 – HDA, which determines its quality ($>5\%$ is preferred).

1.5.2 Uses of Royal Jelly

Royal jelly is recommended for stomach, liver and digestion problems, high blood pressure, loss of appetite, weight loss, fatigue, listlessness, insomnia, pregnancy, menopause, old-age problems, convalescence and athletics. Royal jelly can be viewed as a tonic to make you feel stronger, healthier and less tired. It can be eaten pure or mixed with honey. It is also often sold in glass tubes or capsules mixed with sorbitol or another sweetener. In many countries, it is also added to energy drinks. Capsules with dried royal jelly are commonly used in apitherapy. For external use, royal jelly is added to creams and salves, because it enhances or preserves the beauty of the skin. It stimulates the formation of healthy tissue and hair growth.

1.6 Bee Venom

It is used by the bees as a defensive weapon to protect their property. Nature provides the honeybee with this venom. Otherwise, insects, some birds and reptiles would not allow them to enjoy the fruit of their labour. The African bee is aggressive and stings painfully, and this serves it well, for otherwise human beings, too, the worst enemies of the insect, would rob them easily. The venom has two main medical uses: as a desensitizer for those who are allergic to bee stings, and in the treatment of arthritis. It is applied directly or by inject ion (FAO, 1990).

Uses of venom

In traditional medicine in Africa finely ground bees were used as a salve or tea to combat various diseases including rheumatism. People also had themselves stung on specific places on their body. Bee venom is used in various ways: it is inhaled, eaten in the form of bee venom honey, injected in the form of injection fluid or applied on the skin as a salve. It is also applied by being stung, either on its own or in combination with electrotherapy, acupuncture or acupressure.

This is very painful, and it can be dangerous. In China and Japan, only the removed stinger is used as a needle on acupuncture points. This is felt by the patient, but it is not painful. A minimal amount of bee venom is naturally present in honey. It is of course also present in the mother tincture *Apis*, which is used in homeopathy and natural medicine.

2 HARVEST AND PROCESS HONEY

2.1 Honey Harvesting Seasons, Indicators and Requirement

2.1.1 Honey Harvesting Seasons

Honey is best harvested after the peak of the bee season. The quality of honey changes during its production in the hive, so the selection of which combs to harvest determines in part the quality of the honey. All extraction is best done right after harvesting when the honey is still fluid. When removing combs from the hive, application of too much smoke should be avoided. Honey in freshly built combs can be packed and sold right away as cut comb honey, without extraction or processing. It is important to separate combs before extraction, and harvesting of full combs is preferred. It is better not to harvest combs that contain unripe honey, bee bread and brood if pure honey with a low moisture content is desired. Separating combs with different honey colors and extracting the honey separately will enable a beekeeper to diversify his or her production. Honey in freshly built combs is often lighter in taste and colour (Marieke et al, 2005). The time to harvest honey depends on the flowering period of the bee forage plants and the extent of the honey flow.

Most traditional beekeepers identify honey harvesting season by the experiences they developed in their respective area. The different indicators that the beekeepers use for identifying honey season are: smelling of honey, accumulation of bees around the entrance of hives, end of flowering season and weighing of the hive. Some beekeepers identify honey season by inserting a thin size stick in to the hive. If there is honey, the stick comes back with the honey strips. This method of indicators could not be efficient in identification of honey from brood by weighing and it is also impossible to identify externally whether the honey has ripened or not. But in the case of movable comb and frame hive, the maturity of honey and pure honey can be easily observed. Honey is harvested in the study area from October to December and May (peak periods) and sometimes January to March in each year. Most respondents (78%) harvested honey twice within this period of the year, where as 16% of the sample farmers harvested once in a year and 6% of the sample farmers respond that they harvest three times in the same period, which indicates the presence of high potentiality of the area. It was reported that any production obtained in the remaining periods of the year would be left as food for the colony to strengthen it for the next year. When the honey is ready to harvest, it must be removed immediately. Honey can be considered ripe when 70% of the comb is sealed harvest (Chala et al., 2011).

2.1.2 Indications for Honey Harvesting

- There is strong aroma of honey smelling.
- Clustered bees around the hive entrance
- Bees become idle or less traffic at hive entrance.
- Consider the calendar of the area from the previous observations
- Finally, open and check for the ripe and sealed honey combs.

Early cropping (honey harvesting) is important: To force the bees to collect second round honey either for their own or for second harvest. It may avoid the consumption of honey store by the bees particularly if there is rain. If harvesting is at late flowering, leave some provision to the bees, or if you remove all the honey immediately feed them with sugar.

2.1.3 Honey Harvesting Requirements

Tools and equipment required to remove a honey crop from a hive include:

Bee blower: Blowers consist of a motor-driven impeller that generates a stream of air along a plastic pipe or hose with a restricted outlet.

Smoker is a manually operated material used smoke the hive. It calms down the bees (subdue) & induces them to engine bees full of honey are easier to handle & expel the bees from the surrounding during work.

Bee brush is a material made up of soft sisal fibber mounted on wood and used to remove the honey bees from the honey combs and draw the bees to hive while transferring

Escape boards: The bee escape board is basically a one-way device set into a board that is placed below the super or supers of honey to be cleared of bees.

Hive tool is material made up of iron metal which is sharp on one end. It is used to open the hive, clean propels, wax and unnecessary materials from the frame & hive.

Personal Protective Equipment (PPE)

- Glove: the material used to protect head and fingers from honeybee's sting
- Bee veil: is the material used to protect head region, face & neck from bee's sting.
- Overall suit: made up of cloth used to protect the overall body except the above
- Boots is made of plastic used to protect legs from bees.

Other equipment

- Fresh water
- Loading equipment
- Means of transport for honey-filled frames to extracting facility
- Spare boxes
- Tarpaulins or other waterproof coverings

2.2 Honey Harvesting Procedures

1) Wear protective clothes and puff some smoke gently around the hive.

2) Then puff continuously through the main entrance for at least 3 minutes. Wait for bees to rush.

3) Use the hive tools open the lid or cover of the hive & puff some smoke.

4) Insert the hive tool under the corner of super lever it up. Put a piece of wood to support while the other corner is lifted.

5) Puff in a little smoke, gently rise the super give a light twist to free from super.

6) Insert super-clearing board or bee escape, stand by for the bees to clear into brood chamber.

7) Carry the super box & extract honey.

Once harvested, honey need not necessarily require further processing. On a small scale, simple equipment as used in other forms of food preparation is adequate: plastic buckets, bowls, sieves, straining cloths and containers. Honey is a stable commodity with a long shelf life: if harvested carefully and stored in containers with tight-fitting lids, it will remain wholesome for several years. Honey is a food and it must therefore be handled hygienically, and all equipment must be perfectly clean and without any odour of cleaning materials. Honey processing is inevitably a sticky operation. However, because honey is hygroscopic and will absorb moisture, all honey processing equipment and containers must be completely dry. Any water being added to honey increases the chances of fermentation.

There are some rules which you should keep in mind when harvesting honey:

- Only remove combs with capped honey; uncapped honey contains too much water and will start to ferment.
- Do not take any honeycombs containing brood. In the fixed comb hives, only take

away the combs at the sides of the hive. In hives with loose frames, only take the frames out of the honey area and, at the most, the side frames out of the brood chamber.
- ➤ Sometimes more is paid for certain kinds of pure honey than for a mixture of different kinds of honey and it is then worthwhile harvesting this kind of honey separately (P. Segeren, 2004).

2.3 Honey Extraction, Processing and Packaging

Harvesting bee products, the beekeeper extracts the fresh, primary bee products. Because of their freshness, these products have the highest value for therapeutic applications. For consumption, preservation and marketing purposes, the beekeeper processes the products further, which usually (but not always) increases their market value. Honey is removed from the comb, where by the honey and wax are separated. This is called extraction. The honey is then put into jars and the pure wax is extracted from the empty comb. This wax is worth more and is less perishable than crude wax, but the honey in jars is worth less than well-produced fresh honeycomb (Marieke et al, 2005).

Cut-comb honey: Because the whole comb is harvested from these hives, it is possible to harvest pieces of cut-comb honey for sale this way. Select pieces of comb consisting only of sealed and undamaged honey comb, cut them into neat portions and package them carefully for sale. Since the honey in the comb is untouched and is readily seen to be pure, honey presented in this way always fetches a good price, and honey that has not been open to the air has a finer flavour than honey that has been subjected to processing in any way. Beekeeping equipment suppliers sell cutters to cut uniform sizes of comb, and plastic boxes with transparent lids for selling cut comb honey. The sharp edge of a tin can make a useful comb cutter.

Strained honey: Because combs from fixed comb hives or movable comb (top-bar) hives do not have the support of a wooden frame or wired foundation, they would break up in the type of extractor used for frame hives and the product would be a mixture of honey and fragments of wax. The simplest way to prepare strained honey is to remove the wax capping of the honeycomb with a knife, break the combs into pieces, and strain the honey from the wax. Make sure that you do not use unsealed combs containing unripe honey or pollen. Strained honey must not contain any trace of beeswax or other debris. It is best to first use a coarse strainer to remove large particles, and then to use successively finer strainers. Use a cotton cloth, basket, or sieve to strain the honey from the pieces of honeycomb. Collect the honey that strains through in a clean and dry container. Finally squeeze the combs inside a bag made from the cloth to remove as much honey as possible. Do not discard the empty wax comb, it is valuable!

Honey can be separated from the comb in various ways: through floating or dripping, pressing, or centrifugal extraction. The floating and dripping methods make use of differences in density. In the floating method the wax floats to the surface and in the dripping method the honey drips from the comb. Dripping, floating and hand pressing honey combs are traditional beekeeping methods, but if practiced well they can be very effective and give good honey. Dripping and floating will often lead to a higher moisture content, especially in the rainy season. Before pressing, combs are wrapped with mesh material to retain the wax particles. The honey extracted in this way is less clear than with dripping or centrifuging. Plastic screening material and stainless-steel sieves are better than cloth as they are more hygienic and leave no (cloth) particles behind that may serve as kernels for crystallization.

Honey should be processed in a closed environment where bees and other insects cannot enter. All ventilation openings must be screened with fine wire mesh. Using a tent well sealed at ground level, is an option. Best for quality is to extract the honey as soon as possible after collecting the combs from the hive. If it is, however, necessary to store the combs, it should be done in a well-sealed container. Sometimes the water content in capped honey is too high. You may dehydrate it a bit by placing the honey supers, after uncapping the combs, on top of each other and blowing dry air from below up through the combs. The honey will lose moisture but more importantly, it will lose some of the fragrance.

2.3.1 Honey Extraction

Centrifugal extraction of honey

Centrifugal extraction using a centrifugal honey extractor is a good method for movable combs from chambered hives or top-bar hives. The advantage of centrifuging is that you can extract the honey very quickly and you can use the combs again. Requirements for centrifuging honey are: honey extractor, uncapping knives or forks, one or two basins 15cm deep made of aluminum, tin, galvanized iron or plastic, in which a few uncapped frames can stand, an uncapping tray, a honey strainer or nylon stocking, cheese cloth and a vessel. It is a machine used to extract honey from combs and framed combs. It comes readymade. Some are made of food grade plastic while others made of food grade stainless steel. They have extracting capacity ranging from 2 frames to 18. The combs or frames are arranged either radically, triangularly or rectangular to extract honey. There are manual extractors as well as the electrical ones. All types have a spout for draining the honey out of the tank. The bottom is convex inside to allow all the honey to drain. They are fitted on 3 stands. The main body is cylindrical. They have 2 transparent plastic covers.

1) Uncapping honeycombs

Before extracting the honey, the capped cells in the comb should be uncapped. This can be done with an uncapping fork or knife. Uncapping with a fork is more accurate but slower than uncapping with a knife. Uncapping knives must be filed well and be razor sharp. At temperatures below 25℃ you can make uncapping easier by preheating the uncapping knife in hot water. With a preheated knife, try to cut away the cell capping in one movement.

Irregularities in the comb will result in some of the cells remaining capped. It is best to uncap over an uncapping tray with a wooden bar on which the frame rests. You can leave the wax capping to drain and later melt the wax capping to obtain first-grade wax. The centrifugal honey extractor consists of a cylindrical kettle in which a square or triangular cage made of a frame covered with strong wire mesh turns on an axle. A radial extractor is different in construction because of the radial placement of combs, but in principle its function is of course the same. In the honey extractors two or more frames can be extracted at the same time. The cage in which the frames are placed is turned by means of a handle with gears.

2) Positioning the combs in the centrifugal extractor

All combs must be uncapped before you centrifuge the frames. Place the combs inside the square cage against the wire mesh sides. Turn in the direction of the bottom bar (as the cells face the top bar). Make sure you turn the handle slowly; otherwise the weight of the honey inside the comb will press the comb through the wire mesh of the cage. Gyrate the cage till about half of the honey has been centrifuged from the exposed side. Then position the frames in reverse and turn the handle until the cells on this side of the combs are completely empty. Finally turn the combs once again and turn the handle until the cells on the first side of the combs are also completely empty. Empty combs are put into a honey super and given back to the colony, so that they can lick the cells clean. If the honey flow stops, the empty combs must be removed. A honey extractor is a machine to remove honey from combs in frames by rotating them at high speed so that honey is thrown out of the comb on to the wall of the extractor, and then runs down to the bottom of the drum. Honeycomb built inside a wooden frame is not damaged by this process and when empty, can be returned to the hive.

Honey should always be strained as it runs out of the extractor so that any pieces of wax capping, dead bees or splinters of wood (from frames) are removed. Extract combs of honey taken from hives immediately. All beekeepers need to know how to extract honey from their combs. Before extracting your honey, you need to consider the following: where you will be extracting honey, the equipment you need, the timing of the extraction and quality assurance, food safety requirements occupational health and safety and other special requirements, for example, for organic honey. The electrically heated knife is fitted with a heater element incorporated in the knife blade and can be used continuously, provided power is available. Although this knife is the most expensive option, in practical terms it is by far the best. The steam heated knife is also designed for continuous use. A copper tube soldered to the back of the blade of the knife is connected to a steam source with a flexible rubber or plastic tube. The steam leaves the knife through a second flexible tube and is condensed in a container of cold water. A steam boiler is required for this knife, but suitable approved boilers were not available for hobby beekeepers at the time of writing. Larger boilers suitable for sideline and commercial beekeepers are available, but the beekeeper must determine what regulations apply to their use in the relevant state or territory.

When uncapping with a hand-held knife, rest the frame in a vertical position on an upward projecting stainless steel screw or spike protruding from a piece of wood that spans the gap over the capping receptacle (Rural Industries Research and Development Corporation, 2014). The spike allows the frame to be rotated and reversed without lifting it. Grip the frame so that the thumb is lying along the end bar which is now uppermost. Keep the beveled edge of the knife towards the comb. Uncap a small part of the upper end of the comb with an upward cut. Next, uncap the whole side of the comb with a downward, slightly sawing motion. The depth of cut is guided by the top and bottom bars of the frame. Tilt the vertical comb slightly forward so that the cappings drop away from the comb surface as they are shaved off. Turn the frame and uncap the other side of the comb.

Comb scratcher: Combs should always be cut back to the top and bottom bars of the frame as this keeps them an even thickness. Any patches of caps not removed can be cut away using the tip of the blade or a capping scratcher designed for the purpose. Avoid uncapping any patches of sealed brood. Ideally, combs containing brood should not be taken from hives for extracting. Use the uncapping knife, or a clean paint scraper, to remove burr comb on the top and bottom bars of the frame so it can fit into the extractor basket. The capping scratcher is a small hand tool with multiple long prongs used to 'scratch' open cappings of honey cells. These are available from bee goods suppliers or a good home mad version can be made from a shearer's comb. The scratcher is used by some beekeepers as an alternative to the uncapping knife. The caps are not totally removed, but the bees will later remove them and repair the comb. The scratcher will result in far more capping in the extractor and strainer than when a clean cut is made with an uncapping knife. Uncapped combs waiting to be extracted are best placed in a hive box sitting on a drip tray. This will help to keep the floor of the extracting area clean. They are essential if supers of combs are to be taken inside the house (Fig. 21).

Pressing the honey

Scrape open the combs, break them into pieces and tie them up in a clean cloth (cheese cloth, sheet, pillow case). Knead the combs in the cloth and then press the honey through the cloth.

You can wring out the cloth (you need two people for this, or one person and a fixed point), but it is faster to work with a wooden press. Pour the honey through a clean cloth or sieve into a pot or maturing vessel and leave it to stand for a few days. Any remaining wax particles and pollen grains will float to the top and can be skimmed off. Then pour the honey into a storage jar (airtight). If you want to be able to fill small pots and jars easily, you could use a container with a tap.

Floating the wax

Remove the wax caps of the capped honey cells with an uncapping fork or knife. At temperatures of less than 25℃ you can make uncapping easier by holding the uncapping tool in a basin of hot water for a short while (but do dry it before use). The combs are broken into small pieces and placed in a pot or other container. The container is sealed to make it airtight. After a few days the wax which has floated to the top can be skimmed off. The honey is strained through a clean cloth, nylon stocking or special honey sieve and is again put away for a day. Any foam and wax particles which have floated to the surface can be skimmed off and the honey can then be put into jars.

2.3.2 Honey Processing

Processing of honey is necessary to prevent fermentation by presence of yeast and to retard granulation to kill yeast, heat indirectly in water bath.

- Heat the honey 55 – 59℃ using 40 – 50 – micron filtering mesh by this wax & pollen is filtered.
- Then heat up to 60 – 65℃ 35 and 20 minutes respectively to kill yeast and filter.
- Pass the honey to cooling and settling tank, for storing honey always use steel containers with lid.

Honey is classified by the source from which the bees gathered the nectar, because the source influences the honey's flavor, color and viscosity (thickness). For example, honey collected between October and December may have an orange flavor, showing that the bees collected most of the nectar from citrus trees. The following table shows some plants, with the color and viscosity of the honey they produce (Food and Agriculture Organization of the United Nations Rome, 1990) (Table 8.2).

Table 8.2 Some plants, with the color and viscosity of the honey they produce

Plant	Colour	Viscosity
Orange	Light, yellowish	High
Neem	Light, amber	Very low (runs like water)
Coconut	Dark	Moderate

In general, honey of high viscosity flows slowly. The harmattan wind has a great influence on honeys collected between December and March in West Africa, causing them to thicken. The coastal areas are less influenced by the harmattan and therefore produce honeys of lower viscosity. Heat is applied to force honey to flow; however, the beekeeper must not subject honey to higher temperatures than 30℃.

2.3.3 Honey Packaging

Honey can be packaged raw. Fresh honey has the aroma of the flowers the nectar was collected from. The content of biologically active substances such as enzymes is highest in fresh and unheated honey. The honey should be packed in clean dry glass container with wide mouth slightly and then stirring it. Creamed honey made of fine crystallized honey tastes the best. Within a few days after extraction, pour the honey into an airtight storage jar or containers. To fill small pots and jars easily, use a container with a valve. Store honey in glass jars or plastic buckets with well-sealing lids or in metal containers that have been coated on the inside with liquid paraffin or plastic, or that have been treated with food-safe varnish.

Large honey companies warm the honey to keep or make it fluid and to prevent fermentation if the moisture content is too high. After heating, the honey is filtered and poured into glass jars. This process is also called refining. Through the heating process, however, the honey loses some of its quality. Its fresh character is gone but it does stay clear longer. This is an advantage if the honey is to be sold in stores.

2.3.4 Honey Storing

The commonly used traditional storage containers by beekeepers are clay pot, and container made of cucumber ('kil'). The containers are fitted with lids made of locally available materials and sealed with mud and ash mixture. Such traditional containers will absorb moistures or may change the flavor of honey and deteriorate the quality of the honey. In addition, these containers can break down easily, so they need careful handling and nowadays they are gradually replaced by tin and plastic containers with fitted lids (Chala et al., 2011). Store the honey in glass jars or plastic buckets with well-sealing lids or in metal containers which have been coated on the inside with a layer of liquid paraffin or plastic, or which have been treated with acid-resistant varnish. In humid areas, the honey must be stored in airtight containers within a few days after extraction.

The honey can start to ferment during storage if the water content is too high ($>19\%$). Fermentation can be prevented by heating the honey to a temperature of 55 or 60℃ over a period of 8 hours, followed by rapid cooling. However, heating the honey for too long will cause both the taste and smell of the honey to deteriorate. Heated honey is of an inferior quality as the enzymes are broken down. Honey is best stored in clean, dry buckets with tight fitting lids. If it is kept away from heat, it can be stored this way until it is packaged for consumption or sale.

Honey is hydroscopic, that is, it will absorb moisture from the atmosphere or damp surfaces that it meets. It is the moisture in honey that causes fermentation to begin. This can seriously affect the quality and longevity of your product. Laboratories and very large commercial operations may use a refract meter to determine the moisture content of honey. This is not practical for most beekeepers although some beekeepers do use them. A simple way to test the density of honey and therefore estimate the moisture content of your honey is to place the honey in a jar, leaving a small amount of air and put the lid on it. Turn the jar upside down. The longer it takes for the bubble to rise to the 'top', the denser the honey and the lower the moisture content (Australian Honey Bee Industry Council, 2008).

3 PROCESS OLD AND BROKEN COMBS IN TO CLEAN WAX

Beeswax melts at temperatures between 63 and 65℃. Over heated or burnt wax is worthless. Wax that has been melted in iron, copper or zinc pots loses its smell and colour. Slow heating over a long period has the same effect. The quality of beeswax is judged from its colour and purity. Light wax has the highest value. Old combs are dark and must be processed separately, but they yield very little wax and are hardly worth the trouble. The wax must be washed before it is melted. Put the combs in clean water. Do not use tap water if you know that it is very alkaline, or add a tablespoon of vinegar to every liter of water. The wax must be melted down to purify it further. Among the various possible methods for melting beeswax the straining method and the solar wax extractor are described below:

3.1 Straining Method of Wax Processing

Place the combs and cell capping in clean water and leave them to soak for several hours so that any remaining honey is dissolved. Use an aluminum or plastic basin or bucket. Tie up the wax in a clean cloth. Put this in twice its volume of clean water and heat this until the wax has melted (70 - 80℃). Do not let the water boil! The wax should not touch the bottom of the pot, so you should jam a piece of wood in the bottom of the pot. Place a heavy object on top of the bag of wax so that it remains submerged. The wax will filter through the cloth and float to the top. Pour the warm water and wax mixture through a strainer or a clean cloth and use two sticks or something like squeeze the last of the wax out of the first cloth. Leave the water and wax mixture to cool. If you have first rinsed the pot with soapy water, you will be able to remove the wax cake without any further trouble. If not, cut the wax cake loose from the sides of the pot. To obtain even purer wax, you must first let the wax set, scrape off the dirt at the bottom of the wax cake and then melt the wax again. To do this uses a double boiler or a pot floating in a second pot of water. Then you can again scrape the impurities from the bottom of the wax cake.

Procedure of extraction with boiling water and a wax press

1) Keep wax from comb harvested for honey separate from the dark or old combs that have contained bee brood.
2) Wash the crushed honeycombs in water until they are free of dirt and honey.
3) Put them into a suitable cloth sack and tie with string.
4) Heat a good quantity of water in an old cooking pot.
5) Put in the sack of honeycomb and push it down under the water.
6) Clean comb from harvested honey makes the best wax (Fig. 22).

3.2 Solar Wax Extraction Method

The solar wax extractor provides a simple and effective way of melting and purifying beeswax. It uses the sun's heat to melt the wax, and an effective solar wax extractor can be easily 'home-made'. The temperature inside the extractor needs to rise only to 68–70℃ to melt the beeswax sufficiently: if clean wax is used, just one melting in a solar wax extractor can produce a satisfactory block of top quality wax. The solar wax extractor consists of a glass or clear plastic lidded box containing a sloped sheet of metal. Pieces of honeycomb are placed on the metal sheet and as they melt, wax runs down the metal slope to a container. The sheet of metal can be bent at the edges to funnel wax towards the container. Solar wax extractor/uncapping tray consists of:

- Two basins which fit on top of each other and a lid made of a wooden frame and two sheets of glass or transparent plastic with a space of 5–10mm between them.
- The top tray is 20cm high and wide enough for a frame from the honey super to be suspended by its handles from the sides of the tray.
- A screen of wire mesh prevents pieces of comb and debris from slipping down into the container. Impurities in the wax tend to remain on the metal, and others can be scraped off the final solidified block of wax.
- The bottom is made of fine wire mesh. In this tray, you put the frames or the cell capping which are to be melted.
- The bottom tray has sides about 7cm high and a bottom made of zinc.
- This tray collects the molten wax. To ensure that the trays and the lid do not move apart you should fix protruding metal corners on two sides of both trays.
- To obtain a high temperature in the solar wax extractor, you should paint it black and make sure that is closes well.
- Prop the solar wax extractor at a slight angle, facing the sun, so that the wax runs to one side.
- The wax can be melted again in hot wax uncapping tray so that it can be formed into cakes. Frames that have been uncapped on both sides can be placed against the sidewall of the top tray until they can be put into the centrifugal honey extractor. You can melt the wax once again to purify it further.
- To retain its smell, it should be wrapped in plastic and kept in a cool dark place. Always check your storage for the presence of wax moth (Fig. 8.1).

3.3 Steam Wax Extraction

This device produces steam from a separate boiling pot. The steam is guided with a valve to a perforated sieve or bag that is attached in the wax melting chamber. The wax thus drips to the bottom and is tapped with a valve. The steam master can process large amounts of comb efficiently and it is suitable for all sources of wax. But it is difficult to fabricate such

Fig. 8.1 Solar wax extractor
Source: Pam Gregor, 2011.

a wax steamer on your own; which can be done with the sun and hot water wax melters.

Another type of steam wax melter is heated from below, e.g. by fire. Steam ascends from boiling water in a lower part, underneath a perforated holder full of comb or wax capping. The wax in the upper part melts out of the comb and drips into a catch basin that can be taken out separately.

The wax parts can be scraped from the bottom part once they have cooled. Double-walled steam wax melters heat wax in a central body through a double wall that is surrounded by steam. Steam extractors all work on the same principle: two connected tanks are fixed, one inside the other or one on top of another. The combs or capping are put in an openwork metal basket inside the main tank.

Steam extraction is a good method for capping but is less suitable for melting down old combs as it yields only around 80 percent of the wax. Cleaning the wax in the ways described above will satisfy most wax users. If very pure bees wax is required for special purposes the wax must be refined. Wax can also be cleaned using steam. A sack of wax capping or comb is hung above a metal container floating on boiling water. Melting wax drips from the bag to collect in the container. Take care the water does not boil dry.

4 COLLECT AND STORE POLLEN GRAIN, PROPOLIS, BEE VENOM AND ROYAL JELLY

4.1 Pollen Collection and Storage

Honey bees collect pollen from the stamens of flowers. The pollen sticks to the bee's hairs while the bee is sucking nectar. The bee removes the pollen from its hairs using a comb on its forelegs and adds some saliva to help roll it into a ball. The bee flies with these loads in the pollen baskets on its hind legs to the beehive. Bees push these loads with their heads into the honeycomb cells, together with a small amount of honey and saliva. The bees then

MODULE 8
HARVEST, PROCESS AND STORE BEE PRODUCTS

process this mixture and it ripens into bee bread.

Production of pollen is only possible in the early part of a season, in an area with good vegetation made up of pollen-rich plants and with a strong colony. Harvesting pollen is not good for the development of the colony because the colony may not have enough pollen left to make bee bread and bee milk, which are needed to feed to the young bee larvae. Some pollen must therefore be left behind, for example by not harvesting every day and by rotating the production colonies.

4.1.1 Harvesting

Pollen is harvested with the help of a pollen trap. Harvesting/collecting, need no specific skills; by fitting pollen trap across the entrance, it can be harvested. When the bees are collecting a lot of pollen you can use a pollen trap, placed in front of the flight entrance, to collect the pollen pellets. After a few days, move the trap to another hive so that the colonies are not deprived of too much of their protein supply. Trap consists of a wire grid 5mm mesh (3.6 to 4.2mm). The pollen is collected in pollen tray. The collected pollen is dried in sunlight or using air drier and stored (Fig. 23).

This includes a grid that the bees must pass through when they return to the hive. The entrance holes, which can be round or lobed, are so small that the loads are scraped off the bees and fall through a grid. The bees cannot get through the grid to pick them up again. The various colors of pollen loads are all mixed together in the collection drawer.

4.1.2 Storage

Pollen spoils quickly and can therefore be left in front or under the hive for no longer than a day. The loads must be dried immediately after harvesting to prevent mouldiness and extend their shelf-life. The moisture content decreases during drying from about 25% (fresh) to an average of 11%. Fresh pollen becomes mouldy after just one day, and these moulds can produce unhealthy aflatoxins. To keep it longer, fresh pollen can also be added to honey, but the concentration must be no more than 10%. Pollen must be stored in a dry, dark place to retain its good properties. Brown glass jars are better for this purpose than clear glass jars.

4.2 Propolis Collection and Storage

Propolis are also called as bee glue. It is a sticky plant resin which bee gathers from trees and other vegetation. It is used to seal and repair combs by bees. Harvesting propolis/collecting propolis is made by bees out of tree gums, glues, waxes and resins.

These can be found around the flower buds and are excreted as drops from the tree's bark if it is cut or cracked. The bees bring them on their hind legs, just like pollen, to the hive. They mix them with their own wax and saliva. This produces propolis. Propolis collection and storage methods are as follows:

➢ Place a slotted sheet or wood strips 2 – 3mm wide gap at the top of the hive.
➢ The bees will close the slots with propolis.

> Remove the closed slots and keep it in deep freeze, the propolis then becomes brittle and easy to remove.
> Flakes of propolis can be stored for years in plastic bag.

Propolis have antiseptic, antibiotic, antibacterial and anti-fungal properties. Propolis can be used as human medicine, in cosmetics and healing cream, treating digestive disorders and in throat pastilles and chewing gum.

4.3 Royal Jelly Harvesting

The young bees add secretions from glands on their heads to the ingested bee bread to make bee milk or royal jelly. They put this bee milk in cells that contain young larvae. The larvae of worker bees, drones and the egg-laying female (the queen) eat these products, which make them grow. The bee milk is made up of two components: a clear and a milky white fluid. Royal jelly consists of approximately equal parts of these two, whereas the bee milk for the drones and workers is made up mostly of the clear component.

The bees produce the most bee milk when they are a week old; after three weeks the secretions stop, and they go outside to collect nectar and pollen. To produce royal jelly, it is therefore important to have many young bees in the colony.

4.4 Bee Venom Collection and Storage

Female bees, namely the worker bees or the queen, have a stinger on the end of their abdomen that they can extend. The queen usually only uses this to lays eggs, but she can also sting with it. Worker bees do not lay eggs usually, but only sting with it. A drop of fluid, the bee venom, hangs on the extended stinger. The stinger is also covered in barbs. The bee venom is made in the venom gland and is stored in a venom sac at the base of the stinger. Young bees have little venom. Their venom sac is not filled until their 15th to 20th day, when it contains about 0.3mg of liquid venom. The spring bees that are raised with a lot of pollen have the most and most effective venom. Bee venom dissolves in water but not in oil. Alcohol is harmful to bee venom.

Bee venom is a poison and it can kill both humans and animals! For its collection, harvesting and processing special precautions are needed like gloves, a mouth-cap, etc. Do not inhale or consume bee venom in any way without carefully following prescriptions and calculations regarding the dosage.

4.4.1 Production

Bee venom is harvested using a bee venom collector. This is a glass plate over which metal wires are strung that is electrified with one large battery or many small batteries. When the bees touch the wire, they empty their venom sacs. After many bees have released their venom, the colony attacks the collector plate so that thousands of bees empty their venom sacs onto it. The venom dries up on the glass plate and the jelly-like powder can then be scraped off. Protect your hands with gloves to ensure that you do not meet the venom and cover your face with a mask to keep from inhaling it. The bee venom collector is placed in the hive for an hour and is then taken away. During and after use of the collector the hive and other colonies in the area can become very agitated. It is therefore best to do this in an isolated area.

4.4.2 Preparation

It is produced by poison gland in worker bees. Bee venom consists of several biochemical substances like histamine, dopamine, melittin and enzymes phospholipase, alpha glucosidase.

Physical property: Specific gravity is 1.13. It is a clear liquid and volatile.

Harvesting/collecting: Bee venom is harvested by an electrical stimulation lightly; bees are forced to walk through a glass plate. The electric current uses to stimulate bees to sting; the venom dries and scraped off using razor blade.

To ensure exact concentrations, bee venom is added to honey in stages. For example, 0.1g of bee venom is added to 1kg of honey and then 100g of this mixture is again added to 1kg of honey. This gives a concentration of 0.01mg of bee venom per gram of honey.

>>> SELF-CHECK QUESTIONS

Part 1. Multiple choice.

1. _____ is a material used to decap the cells of sealed honeycomb before the frame combs are placed in the honey extractor.
   A. Honey extractor   B. Uncapping fork
   C. Honey sieve   D. Bee brush
2. One of the following is the primary product of beehive.
   A. Pollen   B. Honey   C. Wax   D. Royal jelly
3. _____ is used by the bees to feed the queen bee and the young larvae less than three days old.
   A. Honey   B. Royal jelly   C. Bee milk
   D. All   E. B&C
4. A sweet thick liquid composed of sugars made from nectar.
   A. Propolis   B. Honey   C. Pollen   D. Wax
5. What are the factors affected granulation of honey?
   A. Temperature below 15℃
   B. High concentration of glucose
   C. High available of nuclei
   D. All
6. _____ is a piece of flat metal used for prying the parts of the hive apart and for scraping away the excess propolis and wax.

A. Hive tool B. Uncapping fork
C. Honey extractor D. Queen-cage

7. What are the indicators of harvesting time honey?
   A. Strong aroma of honey smelling
   B. Clustered bees around the hive entrance
   C. Bees become idle or less traffic at hive entrance
   D. Consider the calendar of the area from the previous observations
   E. Open and check for the ripe and sealed honey combs
   F. All

8. All the materials are needed during honey harvesting except _____ .
   A. Smoker B. Hive tool
   C. Bucket D. Water sparer

Part 2. Match the term in the left column with its definition in the right column.

1. Honey   A. Honey processing
2. Honeydew   B. The fluid, viscous substance produced by honey bees
3. Hive tools   C. Honey becomes solid
4. Granulation of honey   D. Used to seal holes inside the hive
5. Pollen   E. High fructose
6. Hygroscopicity   F. Advice used to measure moisture content of honey
7. Propolis   G. Pollen trap
8. Refract meter   H. Honey quality
9. Flavor   I. Produced by bees after they collect
10. Post-harvest handling   J. Metal tool with a flat end, that is used to prize apart the hive

Part 3. Answer the following questions or statements in the space provided or on an additional sheet of paper if necessary.

1. What are the six hive products?
2. What are the steps of honey harvesting?
3. What are the major constituents of honey?
4. Write type of the honey extractor.
5. What are main characteristic of honey?
6. What are functions of honey sieve?
7. How does processing honey?
8. Name the materials and tools required for harvesting of honey.
9. List the materials that are needed for harvesting of honey.
10. Explain the factors occurred of different honey colors.
11. Name the best honey store equipment.

>>> REFERENCES

Kinati C, Tolemariam T, Debele K, 2011. Quality evaluation of honey produced in Gomma District of South Western Ethiopia [J]. Livestock Research for Rural Development, 23 (9).

Mutsaers M, van Blitterswijk H, van't Leven L, et al., 2005. Bee products properties, processing and marketing [M]. Digigrafi, Wageningen, Netherlands.

Segeren P, 2004. Beekeeping in the tropics [M]. 5th ed. Digigrafi, Wageningen, the Netherlands.

MODULE 9:

REARING QUEEN BEES

>>> INTRODUCTION

Queen bee is the head the colony. Unless she is in good condition, it is useless to expect that colony to be productive. To be productive, each colony must be headed by a fertile and vigorous queen bee. She is best judged by the temperament, honey yield, disease resistance and other related traits of her colony. By raising queens, you can change the blood of your stock and improve the performance your bees. Bad traits such as aggressiveness, less productive, less disease resistant, running on the comb when manipulating the colony, swarming, robbing, etc. can be improved by rearing queens from colonies not having these unwanted traits (Hamdan, 1997). Although queen bee can live for four years or more, there comes a point when the she is no longer vigorous and the colony as a whole begins to decline in performance. She is productive usually between one and two year. Studies have shown that colonies with older queen are more likely to swarm than colonies with young queens. Also, young queens are more prolific egg-layers than the older ones (Diana and Alphonse, 1998). Therefore the old queen needs to be removed out from the colony and replaced by the new every one or two years.

The replacement queen can be purchased from queen bee suppliers or be reared by own. In fact, the bees can rear new queens by themselves that you can use to replace your old queens with. However, this isn't, the best way. Because, it is uncontrolled, and you won't know until it happens just how many queen cells you are going to have. Also, the queens may have blood from colonies that have characteristics you don't want to propagate due to their undesired traits, such as a tendency to swarm frequently, aggressiveness etc. You can use queen cells like this in an emergency, but most beekeepers will agree that controlled, planned and simple queen rearing is a far better (Cramp, 2008). In general, raising queens is a great way for a beekeeper to be successful in his/her commercial beekeeping. This module therefore, deals with the basic information and methods of queen bee rearing. The rearing methods, which are set forth in a step by step manner, are written to provide the beekeepers or students a quick and comprehensive technical guide.

This module enables to recognize the natural conditions under which bee colony raise their own queen, identify the critical requirements for queen be rearing, select breeding stock, identify and arrange tools and equipment for queen bee rearing, understand and apply methods queen bee rearing.

1 THE PURPOSE OF QUEEN BEE REARING

Queen rearing is the production of virgin or mated queens for use in the apiary or sale and involves selection of suitable queen mothers and drones to produce bees that possess desirable characteristics. Queen rearing is a specialized process and is an essential part of beekeeping with following purposes:
- Replace old or undesirable queens in colonies.
- Start new colonies/colony multiplication.
- Improve genetic characteristics and productivity of the colony.
- Provide virgin or mated queen to market.
- Replace suddenly lost queen.

2 BASIC REQUIREMENTS FOR QUEEN BEE REARING

2.1 Breeding Stock

Breeding stock or mother colony is the mother stock from which young queens with desired traits are produced. Many behaviors of bee colony are influenced by heritable genetic traits. As the mother of the entire colony, the qualities of a particular queen are expressed in every one of her offspring. These traits can have profound effects on the behavior and health of the whole colony (Zawislak and Burns, 2013). Honeybees are not the same; each race and colony has different traits from each other; these traits depend on selective breeding. If we breed from a colony with bad characteristics, we will be perpetuating these bad characteristics; similarly, if we breed from poor colony, the resulting queen will be inferior (Hamdan, 1997). Therefore, selection of the best performing colonies from which to raise new queens is the first job in queen bee rearing. You might begin the screening and selection process one year before starting rearing. To choose which queens are best, give some sort of test or criteria ratings to a group of hives at one time under similar weather condition, pollen and nectar flow. Here are some of the most desirable traits to look for when selecting colonies (Diana and Alphonse, 1998; Howland, 2009; Zawislak and Burns, 2013).
- *Fertility of the queen*: Queen should exhibit good egg-laying capabilities, at the time of year when workers are needed, when there is available food, not during dearth times.
- *Honey yield*: Some colonies of bees will be better producers of honey than others in the same apiary. Honey production is dependent on outside conditions as well as

colony population, brood production and overall colony health. Typically, strong, healthy colonies are better producers of honey, and therefore, good honey production often indicates good overall colony health. As the beekeeper works to improve other traits that support colony health, honey production should also increase.

➢ *Disease and pest resistance:* Those colonies better able to resist pests and disease should be selected. You can do this as well by identifying those colonies that are the most robust and require the least treatment. You will want to raise queens using stock from these 'star' colonies. Honey bees are very responsive to selection, and in just a few seasons you can influence the overall health of your bees. Bees that exhibit hygienic behaviors are able to detect and remove diseased brood at a very early stage of infection. This behavior greatly reduces the chance that an entire colony will become infected with a contagious pathogen.

➢ *Swarming instinct:* Colonies that swarm or abscond frequently should not be used to rear queens.

➢ *Temperament and gentleness of the colony:* The reaction of a colony when it is approached, opened or otherwise disturbed can be a genetic trait. Gentleness is a great trait for bees to have. No beekeeper wants to be stung. It is a very hereditary trait. You can test a colony for gentleness by vigorously waving a wand with a black leather patch at the end over an open hive. This will alarm the bees, and they may rush up to sting the leather patch. After a minute or so, count the stingers on the patch. The colonies with the fewest stingers in the patch are the most gentle. Gentle strains are especially important when keeping bees in urban settings. Queens should be propagated from calm, gentle colonies, where the bees will not leave the hive due to disturbances by beekeepers or others.

➢ *Long foraging range of worker bees:* Workers should fly greater distances to gather nectar and pollen in areas where there is no flower nearby their hive. Also the colony that store more honey should be selected as breeders.

➢ *Hardiness of the colony:* Those colonies that come through dry, cold, wet or dearth seasons with populous colonies should be used. Winter hardiness is especially important to beekeepers in climates that have long, cold winters. Colonies that survive a long cold spell must be healthy and strong. They must produce and store enough honey to fuel their winter hunker-down. And they should slow down their brood rearing in the fall efficiently, and start up in the spring in time to have their numbers grow to take advantage of the spring nectar flow.

➢ *Growth the colony:* Colonies that bear brood well even without feeding should be selected. Some colonies will adjust their brood rearing to seasonal conditions. They may increase in size prior to a nectar flow, ensuring more foragers to collect nectar. They may also reduce their population during times of dearth period, which allows

them to use stored food more efficiently.
> *Comb building*: Colonies that have a character to build comb in a short period as possible should be selected.
> *Ability of the bees to find the way back to their nest*: Bees those are able to find their way back to the hive without drifting to other hives, should be selected.

2.2 Sufficient Food (Nectar and Pollen)

Successful queen rearing demands suitable conditions. Attempting to rear queens at the wrong time of year will result in poor quality queens. Ideal conditions are a light nectar flow and good supplies of at least three sources of pollen (Johnstone, 2008). A colony used for queen raising must be well supplied with uncapped honey and pollen frames. If there is no nectar and pollen coming into the hive, it should be fed with syrup and pollen to stimulate the production of royal jelly and the secretion of wax, which are needed for feeding and constructing queen cells respectively. The beekeeper must ensure that there is plenty of pollen in the hive, because nurse bees eat pollen to be able to produce royal jelly. In contrast, poorly fed queens will have underdeveloped ovaries and will have a shorter productive life (Hamdan, 1997).

2.3 Sufficient Drone Bees

Sufficient drones to mate with the newly emerged virgin queens are needed. A virgin queen mates approximately with 10 to 15 drones, it has been estimated that a drone population of approximately 30 drones per queen is needed for excellent mating. As indicated by strong hive may have between 300 to 500 drones at peak periods. In order to have the colonies chosen for queen rearing well forward, and the right drones flying in time, they should be stimulated by slow feeding (Hopkins, 1911). Stimulative feeding of 1 : 1 syrup during absence of sufficient forage will help colonies raise drones early (Hamdan, 1997).

2.4 Suitable Weather

Suitable weather for mating of queens and drones is required. Both queens and drones are stimulated to fly on a sunny day with a temperature about 21 to 27℃. During a rainy day or unfavorable weather queens will not fly out for mating (Hamdan, 1997).

3 CONDITIONS UNDER WHICH BEES RAISE THEIR OWN QUEENS

Naturally honey bees rear new queens in response to three conditions: 1) the colony is making preparations for swarming, 2) the queen's physiological and behavioral activities are substandard (when queen fails to produce sufficient pheromones and lay lees eggs), and 3) the queen is lost or dies. In each case the purpose is to replace the existing or lost queen in the colony (Diana and Alphonse, 1998).

3.1 During Swarm Preparation

Swarming is a natural phenomenon and is the bee colony's method of reproduction in which the colony divides into two small colonies each with their own queen. This natural process is known as reproductive swarming. As indicated by Hamdan (1997), the bee colony build about 6 to 20 queen cells on the sides or along the bottom of the comb when preparing to swarm. Then the existing queen lays her fertile eggs in these cells where they are reared into adult queens by worker bees. The occurrence of this natural process varies generally based on availability of the flora, the strength of the hive and the swarming tendency of the bees (Fig. 24).

3.2 When an Old Queen Fails to Lay Eggs

Another condition that leads to queen cell construction without swarming occurs when the bees prepare to replace a queen that is substandard, which is called *supersedure*. When the queen becomes too old or infertile or has a physical mishap or disease, the bees decide to raise a new and more efficient queen to replace her. Workers begin this process by either constructing queen cups in which the original queen lays eggs or modifying existing worker cells containing young larvae. These supersedure cells are few in number (one to three) and found in the center of the comb. After the daughter queen hatches, she may often coexist with her ailing mother. In time, however, only the replacement queen will be found (Hamdan, 1997; Diana and Alphonse, 1998).

3.3 When a Colony Lost Its Queen

The last condition of queen replacement is when the queen is absent from the colony. This can be from natural case, beekeeper error, or predation. Occasionally the queen will fall off the comb during hive inspection and is unable to return to the hive or crushed between two frames as they are being removed or replaced (Diana and Alphonse, 1998). The bees recognize her absence and begin quickly to raise a new queen from worker eggs or young larvae by enlarging and extending the worker cells and feeding royal jelly to the selected larvae. Emergency queen cells are distinguishable from the queen cells of supersedure or swarming by being raised in enlarged worker cells on the comb face and are often smaller in size than queen cells raised from queen cups (Hamdan, 1997).

4 METHODS OF QUEEN BEE REARING

Generally, the methods by which the beekeepers rear queen bees are broadly categorized into two as grafting method (transferring of young worker larvae into artificial cells where they develop into mature queen) and non-grafting methods (Diana and Alphonse, 1998).

Whether the beekeeper uses grafting or non-grafting method, he/she is dependent upon the bees to bring the young larvae to mature queen bees. To ensure this, he/she takes advantage of the natural instinct of the bees, under which they raise queen bees by their own (Hopkins, 1911). Therefore, rearing queen bees artificially, needs forcing the colony to rear new queens by creating one of the above three conditions discussed under which queen bees are reared naturally. Swarm and supersedure impulse are preferred, because they produce high quality queens.

4.1 Grafting (Doolittle) Method

Grafting method of queen bee rearing is named after G. M. Doolittle, a beekeeper who study intensively and wrote about the subject. It involves transferring of worker larvae at 12 – 36 hours old (less than 3 days old), using a grafting tool, into plastic cell cups where it develops into adult queen bee. This is the standard method for producing large numbers of queens (Diana and Alphonse, 1998).

4.1.1 Tools and Equipment Needed

Grafting larva requires specialized tools and equipment (Johnstone, 2008). All items can be purchased from beekeeping suppliers or can be modified from existing beekeeping equipment. Before beginning grafting, prepare the work place by arranging all tools and necessary equipment within easy reach.

1) Queen cell cups

For each larva to be transferred, you will need a queen cell cup. It holds the larvae in a vertical orientation in the hive, which encourages the worker bees to rear them into new queens. Cell cups can be purchased or made by from bee wax. To make your own queen cups, you need to have pure beeswax which is heated to the melting point in a container and a specially designed dipping stick. The dipping stick can be made from a 3/8 inch dowel and the tip is then filed and sanded to form the bottom of a cell and slightly tapered toward the base. The taper is made starting about 1/2 inch from the end of the dowel. The rounded base should be approximately 5/16 inches in diameter (Stahlman, 2007) (Fig. 25).

Once the dipping stick made in such manner is obtained, it is deepened in to molten wax to the depth of approximately equal to depth of natural queen cell and then removed off after dipping in cool water with as little pressure as possible, by gentle twisting motion. The dipping stick should be first soaked in cool water for some minutes to make taking off the cell off it easy and without breakage. The cell cups which are made of wax usually set into a wood cell cup which allows the safe handling of the queen cell (Stahlman, 2007). Becoming very popular and just as good as the natural beeswax, queen cell cups are various designs made of plastic. These cell cups will fit into a slot in the top bar or grooved bottom bar.

2) Grafting tool

Grafting tool is a tool is used to pick up an individual larva each floating on its little raft of jelly, undisturbed from a comb and transfer (place) it in to a queen cell cup. There are different kinds of grafting tools are available which include: carved duck or goose feather, metal grafting needle, 00 point brush, Chinese grafting tool, finely carved matchstick (Johnstone, 2008). Some have a magnifying glass fitted to the stem which can help if one's eyesight is insufficient. Usually both ends are designed for grafting; each offers a different configuration. A very small (size No. 000 or 00) artist's paint brush is a suitable tool for grafting. The moistened bristles must stick together to easily slide under a larva. A 'Chinese' grafting tool is a handy and inexpensive grafting tool that looks like a ball point pen. It consists of a spring loaded bamboo plunger that slides along a thin tongue of flexible plastic. The flexible tongue slips easily under a larva and then a press on the plunger will deposit the larva and any royal jelly that was picked up in the cell to be grafted. A non-slip grip in the middle section gives excellent control. Modern versions of this tool have injection moulded plastic parts, which may help with cleanliness (Fig. 26). You have to use the grafting tool that best fits your needs and your technique for handling larvae (Johnstone, 2008).

3) Modified frame and cell bar

The cell bar is a bar of wood that can be fitted in to the standard bee hive frame (shortened to fit within the frame) horizontally and used to hold artificially made queen cells in upside down position in the way that the naturally made cells are suspended to the combs in the hive. The plastic or wax cells can be attached to the bars by molten wax. Some plastic cups are designed to fit snugly into a standard grooved bottom bar. Usually twenty cells are attached to each bar (Fig. 27).

4) Table lamp

Bright lighting is important when selecting appropriate larvae. A headlamp or a desk lamp that can be easily moved and adjusted is useful for hands-free operation, but a handheld flashlight is also effective to illuminate the bottoms of the cells.

5) Magnification aid

Transferring small larvae from cell to cell requires excellent close-up vision (good eyesight) and so, for many beekeepers, reading glasses and/or a magnifying glass may be necessary (Cramp, 2008).

6) Feeders

Frame feeders are plastic frames with sides and an open top. They are used to feed your bees with sugar syrup, when required. To do this you simply remove the outer frame in the brood box and replace it with a frame feeder. You then fill the feeder with sugar syrup, remembering to place some material in the feeder such as bits of wood or dried bracken, so that the bees will have a foothold and will not drown in the syrup. There are other feeders available but, in my opinion, the frame feeder is the easiest one to use for the beekeeper with only a few hives (Cramp, 2008).

4.1.2 Technical Procedures of Grafting

1) Preparing cell starter colony

Cell starter colony is purposefully made queenless into which the grafted larvae are put for 24 or 36 hours for acceptance and initial feeding. If a queen is lost or killed, a sudden reduction in the level of queen pheromones in the hive usually triggers the worker bees to build emergency queen cells to rear a replacement queen (David and Mike, 2002). Since starter colony is queenless, it strives to replace its queen by feeding and incubating the young worker larvae available in the hive. As its name implies, the starter colony is used to start rearing process, it is not used until the last day of rearing. This is because of the fact that, the queenless colonies cannot feed the larvae with sufficient royal jelly as compared with those of queen right. Starter colonies must be prepared a day ahead of grafting so that the bees can recognize the absence of their queen and start to look for any option to replace her. You can follow the procedure to prepare cell starter colony as explained by Hamdan (1997).

a) Select a good and strong colony and remove the following combs from the selected colony and place in a nucleus hive:
- One or two combs with as much as possible open brood with eggs and youngest larvae, making sure that the queen is not on any of them
- Two combs containing honey and pollen
- Additional nurse bees from at least two combs taken from other hive are shaken into the nucleus.

b) The arrangement of combs in the nucleus hive should be: Open brood combs are in the middle and to the left and right the two combs with honey and pollen grain.

c) Close the entrance of the nucleus with grass and move itto a new position to at least one kilometer to reduce drifting the bees to the original hive.

d) Feed the nucleus colony with 1/2 liter syrup (1 part sugar to 2 parts water) so that the nurse bees are in good condition for queen rising. The feed can be provided with frame feeder (suspended in the hive like a frame) in the left space.

2) Preparing cell finisher colon

Bees naturally rear queens while in a queenright state. A new queen may be reared in a *supersedure* cell to replace a substandard or failing queen. One of factors that induce supersedure is believed to be the poor distribution of queen pheromones among worker bees (David and Mike, 2002). Cell finisher colony is queen right colony forced to feel failing of the existing queen to produce sufficient pheromone and used to finish the rearing of grafted larvae to mature queens. Once the larvae are started to be fed and incubated in the starter colony, a queen right colony can be forced to rear the queen cells to mature cell stage by triggering the *supersedure* impulse, under which bees rear young larvae worker larvae to new queens. Any strong colony will rear started queen larvae. However, when selecting the colonies for cell building, those headed with very young queens should be rejected as

they cannot develop supersedure behavior (Johnston, 2008). Hives suitable for feeder or cell builders are:
- 6 - 8 frames of brood
- Disease free
- Queen is 6 - 12 months old
- Excluder between brood box and super
- Well maintained
- Strong with good tempered bees covering frames in both boxes
- Strong double hives full of bees of all ages and with an abundant supply of fresh pollen and nectar-a double hive is required for each cell feeding colony.

As explained by Johnston (2008), the cell feeding colony should be prepared the day before it is required, in such a manner that:
- The queen confined to the brood chamber with a queen excluder to make the queen doesn't have access to the queen cells as she can kill them.
- Two frames of unsealed larvae 3 - 5 days of age are placed in the centre of the super above the excluder to attract the young nurse bees of a suitable age (5 to 15 days old) days to the super for feeding and incubating the larvae to be put there. If required, support hives may need to be maintained in the same or a different apiary to provide frames of bees, brood, pollen and honey. Fill out the remaining space with combs of honey and pollen. Make sure that combs of unsealed honey and pollen are placed alongside the frames of unsealed larvae. This will simulate a natural brood nest. The unsealed brood should be lifted into the super with the nurse bees adhering to the comb. Quite soon, the bees in the super will recognize the new brood nest is not being occupied by their queen, and begin to respond to the natural supersedure impulse. It is in this environment that freshly grafted or started cells can be introduced for rearing. Brood moved above the excluder should be inspected 2 days after being moved and any self-raised queen cells destroyed.
- A space is left between the two brood combs to make room for the frame holding the bar of grafted larvae.

3) Obtaining the larvae to be grafted

In the colony breeder colony, place a comb in the centre of the brood nest. Eight days later, you will find eggs and newly hatched larvae ready for transfer. These small larvae will usually be situated around the edges of older larvae and capped cells. From this colony, take the comb of eggs and the very young larvae (Cramp, 2008). Grafting is easier from dark wax combs rather than from light wax combs because of the better contrast with the small white larvae.

4) Grafting

Transferring of worker larvae at less than three days old, by the use of a grafting tool into plastic or wax cell cups is called grafting. It is a precise exercise in dexterity and firmness. It takes some period of practice to pick up speed and skill. It requires excellent close-up vision and so, for many beekeepers, reading and/or a magnifying glass may be necessary. The technical steps of grafting larvae are explained by Hamdan (1997), Diana and Alphonse (1998) and Johnston (2008) in detail as follows:

Having made the necessary artificial queen cups mounted on cell holding bars, the usual practice is to wipe the inside of each cell with a little royal jelly produced from another queen cell or placing the bar of the cells in a hive at least 24 hours before grafting so that the bees can clean and condition them. The bars of cells should be prepared by sticking 20 plastic or wax cell cups to per each using molten wax and then fixed to the modified frame.

Next to preparation of queen cups, young (less than three days) larvae containing combs should be collected from a breeder colony, by brushing the bees off the frames rather than shaking to avoid dislodging of the larvae. Then, place the comb onto a support board on the table at an angle of about 30 degrees. Back lighting is important for you to see into the bottom of the worker cells that contain the larva of required age. You must feel comfortable with the table height and the reach to the frame that contains the larva and the cell cup bar (the destination of the larvae). Look for larvae that are so small that you can hardly see them. The ones you want will resemble a small letter c or a comma sitting in the bottom of the cell in a bed of royal jelly. Use a magnifying glass if necessary.

Then, start to pick up the young worker larvae from the comb and place it in the cell cups mounted on the bars by using grafting tool. The grafting tool must follow the curve of the bottom of the cell, so it can be inserted under the back of the tiny floating larvae without touching it. Place the nib or needle of the grafting tool under the middle of the larva and scoop up with some royal jelly (Fig. 9.1 & Fig. 28). It is important for larval survival that each larva is placed in the middle of the bottom of the cell cup. If you experience problems seeing the larva, shave the cell walls down close to the foundation with a sharp knife, then remove the larva. Quickly check your bar when you have finished grafting and re-graft any you think are damaged or too big. The number of grafted cups should not be more than 40 cells/frame. Move your hand along the cells as you graft into them to keep your place on the bar.

The age of grafted larvae plays a major part in the quality of the resulting queens. Younger larvae induce better queens. Select the smallest larvae possible, preferably larvae just hatched from the egg. A larva 0 to 24 hours old is the same length as an egg. Grafting of eggs themselves is extremely difficult. Only graft larvae that are under 36 hours of age from hatching and are floating on a good amount of royal jelly. Grafting of the ideal size larvae is a skill that some people have trouble mastering. The use of commercially available magnifying glass may help where grafting is not required, especially if you have poor eyesight.

After all cell cups are grafted with the larvae, it is necessary to mark cell bar with the date of graft. This is because, regardless of the weather, your schedule, or unexpected

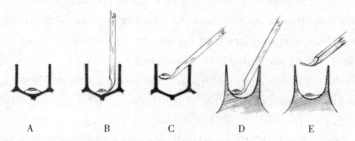

Fig. 9.1 Illustration of grafting a larva from a worker cell to a queen cell cup using a shaped matchstick

A. Larva floating on royal jelly
B. The grafting tool in the vertical position close to the worker cell wall is slid under the floating larva
C. The larva is lifted out
D. The larva is placed into the queen cell by drawing the tool across the cell base
E. Grafting is completed
Source: http://www.depi.vic.gov.au/home

events, your new queens will begin to emerge within 11 and at most 12 days from this day. If just one of the new queens emerges before the others you will be left with just one queen. The identification number of the mother colony should also be recorded to identify from which mother colony the queens are reared.

Grafting may be done indoors under good artificial light, or in the open, if atmospheric conditions are suitable. Don't expose larvae to direct sunlight, and work as quickly as possible. Avoid grafting in very hot weather and/or periods of low humidity as the larvae may be damaged by dehydration. Grafting in the very early morning may help to avoid these conditions while temperatures are still relatively cool.

5) Placing the grafted larvae in to cell starter colony

Once the cell cups are completed with newly transferred larvae, the frame holding the bars must be placed in the cell starter colony immediately. Wrap the grafted bars in a moist towel to prevent dehydration until ready to place bars into holders and into starters (Johnstone, 2008). Because the bees are queenless, they will recognize the cells as 'queen cups' and will draw them out into queen cells. Once the frame is placed into the hive, nurse bees will gather on the frame and begin to feed the young larvae (Stahlman 2007). The bar of cells is to stay there until the next morning. Do not overload the starter colony. Forty cells are quite sufficient. By now, the starter hive has plenty of nurse bees and is fed sufficiently. It is queenless, strongly motivated by the emergency impulse, and has nothing to feed or look after except the bars of freshly grafted larvae. The grafted larvae should be placed between combs of capped brood with pollen and honey in down position. This method of starting cells should not fail barring outside interference from animals or people, excessive temperatures or imperfect grafting. In hot weather, always provide shade for starter colonies, and for that matter, the cell finisher colonies also.

6) Checking the grafted larvae for their acceptance

On the morning of day one after grafting, remove the bars of cells from the starter hive and check for the acceptance of the cell by the bees. The acceptance of cells on each bar will probably vary, depending on different factors, as described in detail by Ruttner (1983). The most important factors are: quality, strength and developmental stage of the nurse bees; age of the grafted larvae; presence or absence of queen in the rearing colony and duration of the queenless stage; presence of open brood in the cell-starting colonies; number of grafted cells; rearing sequence and method of rearing. When checking for the acceptance of the larvae, cell bar holding frame should be always handled gently without shaking or jarring, but can be turned upside down to check the contents of the cells. If accepted, the bees further extend the walls of cells with beeswax, and each larva is floating on a deep bed of royal jelly (David and Mike, 2002). During checking for acceptance, cull any cells whose larvae look a bit bigger than the others and any that do not seem to be well fed or dead (Johnstone, 2008).

7) Transferring the grafted larvae from cell starter colony to finisher colony

On the day (after 24 hours) after grafting, a cell bar frame containing one bar of the started queen cells from the colony the cells were started in is taken and placed between the frames of brood in the super. No more than 20 queen cells should be placed in each finisher colony to ensure that each cell will be well fed (Johnstone, 2008). Carefully open the cell feeder hive and make a space between the two combs of brood in the super. Place one bar of started cells (20 cells) with adhering bees per each colony in this space, being careful not to shake or knock the bar. Replace and refill the feeder. Continue light feeding until the time the cells are sealed, i.e. for 4 days after the cell bars were inserted. Other bars of cells may now be grafted and given to the starter colony. The starter may be re-assembled in its original form if it is not needed to start more cells.

8) Collecting the mature queen cells

The bar of cells is removed from the feeder colony 9 or 10 days after the cells were grafted and the cells have to put in the cage. At this time the cells are ripened and ready to be transferred to nucleus or queen less colonies for mating after they emerged. There has to be a cage for each queen cell since the first emerged queen can kill the rest if kept together. Alternatively queen bank can be used as it has partitions where each cells can kept be individually. Queens will emerge from the cells 1 to 2 days after being placed in the mating nucleus.

Once a cell feeder colony is prepared, a batch of cells can be produced without any further colony management. If successive batches of cells are to be introduced, they can be added after the first batch of 20 cells are sealed, i.e. 5 days after grafting the first batch of cells. Cell feeding colonies in continuous use should be manipulated every 7 days. The queen and older brood are left below the excluder.

4.2 Non Grafting Methods of Queen Rearing

For many beekeepers the idea of grafting and producing their own queens is intimidating. Practically, the methods described below are simple and produce good queens, and are suited to small-scale beekeeper who wishes to obtain a few cells for replacement of undesirable queens and for making their colony multiplication.

4.2.1 Miller Method

The Miller method of queen rearing, named after C. C. Miller, is the easiest for the beginner (Diana and Alphonse, 1998). Miller method is the simplest of all queen-raising methods which does not require special equipment and is suited for a beekeeper wanting to raise a small number of queens, eight or nine queens as indicated by Hamdan (1997) and Howland (2009). The authors explained the technical steps to be followed under this method as follows:

1) Preparing triangular shaped pieces of comb foundation sheet

First triangular shaped pieces of foundation are fastened to the top bar of an empty frame, leaving a space of 5cm at each end. These pieces are 10cm wide and tapered to a point at the end and extend half the depth the frame. The foundation sheet should be well-pegged to the top of the frame with melted wax. Take a deep frame with wax foundation and cut the bottom edge of the foundation into a saw tooth pattern (zigzag fashion) as illustrated in Fig. 9.2.

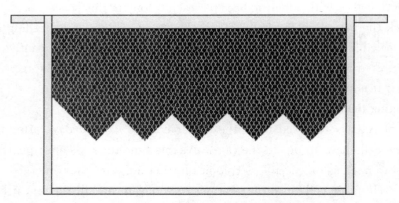

Fig. 9.2 An illustration of zig-zag trimmed foundation
Source: http://website.lineone.net/~dave.cushman/millercomb.gif

2) Inserting the triangular shaped comb foundation in the breeder colony

The prepared frame is then placed in the middle of the brood nest of a colony selected for queen raising (breeder colony) after taking out a comb from the side first to make room. After the foundation sheet is drawn out to full comb by worker bees, the queen will lay her eggs on this comb. In hot weather, it is difficult for the bees to draw out a new foundation and induce the queen to lay eggs on it. Alternatively, a drawn comb, preferably unwired may be used. In order to induce the queen to lay eggs on the foundation sheet/empty comb, there should be no other empty comb where the queen can lay eggs. The hive should be fed if there is no sufficient nectar and pollen at the time.

3) Transferring the comb from breeder colony to cell builder colony

After seven days, the foundation will be drawn out and contains brood with the youngest larvae and eggs toward the margins. The comb is now removed from the breeder queen colony and the bees are brushed off gently while it is held vertically, the comb is not supported by wire and in this case, is fragile to shake. It is important to expose the youngest of the brood or eggs at the edge of the comb (where cuts are made) if they are found to be far from the edge, as the bees prefer to build cells on the edges. The frame is placed on a board or table and with a warm sharp knife the bottom margins of the comb are trimmed so that the cells at the margins contain the youngest larvae (less than three days old). This should be done in a warm, shady place to protect larvae from winds and direct sunshine; the frame should not be out of the hive longer than 10 minutes. This comb should have only small, young larvae. The smallest larvae will produce the best queens as discussed under grafting method. If every second and third larva or egg on the margin of the comb is destroyed, the queen cells will be built farther apart; this will facilitate their easy removal (harvesting of the matured queen cells) in a later stage.

Although it generally takes seven days after insertion of the zigzag cut foundation sheet in the breeder colony, for the larvae to be ready, the colony should be checked after five days as under some conditions the larvae may be ready earlier. Under other conditions, it may take up to ten days before young larvae will be found on the comb.

The prepared comb of larvae is then inserted in the middle of a strong cell-building colony from which the queen is removed (queenless) one to two days beforehand. The bees will build many queen cells along the bottom margins of the comb and the larvae inside will be nursed into queen pupae.

4) Collecting the matured queen cells

After ten days of comb insertion in the cell-building colony, sealed queen cells should be apparent. Take out and carefully brush away the bees. Never shake or turn queen cells on their sides. Harvest the ripened queen cells. Allow room around the cell so that the cutting in no way disturbs the developing queen, and place in queen bank or individual cages for re-queening, selling or new colony establishment (colony multiplication). The new queens will emerge from their cells and soon will mate and start to lay eggs within 10 days. After this, the cell builder colony should be left one or two good queen cells from which to raise its own queen.

4.2.2 Hopkins Method

The technical details of raising queen bees using Hopkins method is explained by Hamdan (1997) as follows. This allows the beekeeper to raise as many as 20 to 30 queens from one frame of brood; however, for good quality queens it is best not to allow bees to produce more than 20 queens at a time.

A comb is taken out of the hive containing the breeding queen from whose progeny it is desired to rear the queens. A new comb of wax preferably unwired is placed in the brood nest. The nurse bees will draw out the cells for the queen to lay in. The comb is removed after 4 – 5 days when it should be filled with eggs and some newly hatched larvae, and the bees are brushed off the comb.

The best side of the comb is selected for the queen cells and is placed flat on a convenient surface. Every 3 rows of cells out of 4 are destroyed either by a matchstick or scored out to the midrib with a sharp warm knife or a razor blade. Then 2 out of 3 cells in these rows are destroyed to the midrib, leaving the third intact as a potential queen cell. This preparation leaves space between these cells in which the bees could easily draw them into queen cells and allows cutting them out. Some practice destroying 4 rows of cells and 3 cells in the row leaving the 4th standing to give more room between the finished queen cells. If the brood frame used without preparation the queen cells will be built in bunches that will be impossible to separate without injury to many of them. The mouths of the cells may be slightly open with rounded-end of a wooden stick.

The comb is now ready to be placed in a strong hive made queenless a day previously to serve as a cell-building hive. The frame is laid flatways above the brood combs of the cell-building hive with the prepared side facing downwards, and raised above the top bars 2 to 4cm by an empty frame or two small blocks of wood to provide space in which the bees could build queen cells. The comb and the top of the hive are covered with cloth or canvas to protect it from cold, and an empty hive or a honey super is placed over.

The bees will perceive this inverted brood to be queen cells because of the orientation, and raise queens of them. When the queen cells are being sealed on day 14, they are removed as soon as possible from the cell-building hive, carefully extracted from the frame with a sharp warm knife and distributed to hives to be requeened or to mating nucs (Fig. 9.3).

Fig. 9.3 Matured queen cells produced using Hopkins method
Source: http://www.ohiohomestead.com/ohioqueenbees/images

4.2.3 Alley Method

In this method, queen cells are raised on a prepared strip of cells. A strip of cells containing one day old larvae is removed from a comb and placed in a frame with the cells pointing downwards. Every 2nd and 3rd larva is destroyed, leaving adequate spacing for queen cells to be started and finished without having to surgically separate the cells once they are sealed (Ruttner, 1983). A comb fitted with new foundation or a drawn comb is placed in the center of a breeder hive for the queen to lay eggs in. When five to seven days later the comb will be found full with eggs and hatching larvae, is removed and the bees are shaken off and laid on a flat surface. Rows of cells containing eggs and larvae not older than three days old are cut in strips from the comb about 12mm wide, using a sharp warm knife. The cell walls on one side are cut down to 6mm of the midrib, and the mouths of cells are little enlarged with a round-ended wooden stick. With a matchstick or small stick, every 2nd and 3rd egg or larva in a row is destroyed, leaving the first cell of three intact. This assures adequate space for queen cells to be built and later to be cut out easily and without injury. If more space is desirable between cells, more eggs or larva can be destroyed between those left intact. Each strip is fastened by melted beeswax or glue to a lower edge of a comb from which 2/3 part has been removed of the bottom, with the selected cells pointing downwards. Alternatively, the prepared strips may be affixed along the underside of the top bar of an empty frame or to one or two wooden crossbars that can be fitted or nailed between the end-bars (Hamdan, 1997).

The frame with prepared cells is given to queenless colony from which the queen is removed. In a day or two, the bees will proceed to fashion the cells that contain young larvae into queen cells. Ten days later, the comb can be taken out and the queen cells can be cut off and given to nuclei prepared for their reception. One cell can be left in the queenless colony.

4.2.4 Splitting Method

In this method, bees are induced to raise queens after removing their queen. The queen cells are raised in the same colony from start to finish. As discussed by Hamdan (1997), Gezhegn and Nuru (2003), the process begins with the selection of the best and strongest hive that has the genetic line that is to be propagated. Then, the colony is split into two or more divisions/parents. To split the colony, remove the queen, two sealed brood combs and a comb of honey and pollen with the adhering bees, and install them in the middle of hive's box filled by an empty comb to form nucleus colony. The entrance of the hive is closed with grass to ensure that sufficient bees stay with the nucleus. The nucleus is moved and placed in other place, and the entrance is re-opened after two days. The bees in the queenless hive will soon become aware that their queen is absent and in due time will start building queen cells in the remainder combs of unsealed brood and eggs.

The original colony (queenless colony) has to be checked after 4 days, and cells touching one another should be thinned out. The cells must be removed a day or two before the queens are due to emerge; this is usually in the 9th or the 10th day after splitting. The small or uncapped cells are discarding, since small cells produce small queens with fewer ovarioles. The good cells are carefully cut out from the comb with a sharp knife and transferred to colonies in which the queens are to be emerged. The nucleus can be returned to the parent colony after the queen cells are finished or harvested and be reunited by the newspaper method (Hamdan, 1997).

It is much better to rear queens by this method during the nectar flow. If the nectar flow is low or there is no nectar flow, the colony must be fed with syrup to encourage bees to secrete wax to build the queen cells.

4.2.5 Overcrowding Method

This method of queen bee rearing is one of the simplest methods that can be applicable at the small holder farmer's level. As indicated by Gezagegn and Nuru (2003), it is widely practiced in Tigray and Amhara regions of Ethiopia. It involves reducing the volume of the hive to induce the colony to prepare for swarming by making it overcrowded. When the bees are overcrowded, naturally they tend be divided in to sub colonies by creating new queens. It is these queens that can be collected and used for the intended purpose before they emerge from their cells. The colony can be made overcrowded either by reducing the super or not providing super when it needs. If the hive is top bar hive, the colony can be overcrowded by reducing the space occupied by the bees using bee barrier which is made to fit the shape and size of the hive. As explained for the others methods, selecting best performing colony (mother colony) is the initial step. After the colony is overcrowded, the worker bees construct queen cells at the bottom and sides of the combs and the existing queen lays her fertile eggs in the cells. Nine or ten days after the colony is overcrowded, sealed queen cells are apparent and ready to be collected for intended purposes. Though, it is not convenient to collect the ripe queen cells from traditional hive, virgin queens can be collected from the swarming groups immediately after they leave the hive.

As the quantity and quality of food that the queen receives during her larval developmental stage is a crucial factor for her growth and strength, queen rearing should be performed during the initial flowering period when the bees can get enough nectar and pollen. If enough feed is not available for the worker bees, they cannot produce sufficient royal jelly to feed the larvae which further influence the growth and strength of the queens going to be emerged. The other point not be forgotten is the time at which the queens can be hatched out from their cells. If the accurate time is missed, the queens may hatch out and leads to colony swarming. Therefore, recording the day at which the colony is made overcrowded and checking after 9 or 10 days is necessary.

MODULE 9
REARING QUEEN BEES

>>> **SELF-CHECK QUESTIONS**

Part 1. Choose the best answer and encircle the letter of your choice.

1. To rear queen bees, breeder stock is selected from:
   A. Aggressive colonies
   B. Swarmed colonies
   C. Productive and healthy stock colonies
   D. Feral colonies

2. One is not an important trait in selecting breeder colonies.
   A. Gentleness
   B. High honey production
   C. Disease and mite resistant
   D. High level of swarming tendency

3. A requirement for rearing good queens is
   A. Five day old larvae
   B. Sufficient nectar and pollen
   B. Sufficient nectar and pollen
   D. Few drones in colony

4. Good honeybee colony growth and well-being is a result of
   A. High egg laying capacity of queen
   B. Few space for the expansion of brood
   C. Less forager bees in the hive
   D. Inability to resist disease and pests

5. Queen bee rearing and its purpose is
   A. To provide old queen to market
   B. To replace suddenly lost workers
   C. For colony multiplication
   D. To replace virgin or new worker bee

6. When to rear queen bees?
   A. Possible at any season of the year
   B. During flowering season
   C. During dry season
   D. All

7. Whether the bee-keeper uses grafting or non-grafting method, his/her success is depend on
   A. The bees to develop the young larvae to mature queen bees
   B. The type of hive used
   C. The race of the colony
   D. The type of food source available for bees

8. Rearing queen bees artificially needs
   A. Feeding the colony with supplementary feeds
   B. The eggs/larvae to be incubated artificially

• 165 •

C. Creating the natural condition under which bees rear queen

D. Inducing the colony to store honey

9. Queen bee rearing method which involves transferring of worker larva from brood comb to where it develops in to adult queen bee is known as _____.

   A. Splitting method   B. Overcrowding method
   C. Alley method   D. Grafting

10. The colony that purposely made queenless in to which the grafted larvae are put for 24 or 36 hours is _____.

    A. Starter colony drone   B. Finisher colony
    C. Mother colony   D. Breeding colony

Part 2. Give the correct answer for the following questions.

1. What is the importance of queen bee rearing?
2. How do you think that queen bee determines the productivity of the whole colony?
3. As a beekeeper/expert, what traits do you desire from bee colony?
4. Compare and contrast the three conditions under which bees rear their own queen naturally.
5. Where and how can you get mother colony for queen bee rearing?
6. What are the advantages of grafting over non-grafting methods of queen bee rearing?
7. List tools and equipment needed for grafting method of queen rearing and explain how to use them.
8. Discuss about starter and finisher colonies.
9. At what developmental age shall the larvae be grafted for production of quality queen bee?
10. Explain the technical steps to be followed in grafting, Miller, Hopkins, splitting and overcrowding methods of queen bee rearing.

>>> REFERENCES

Blackiston H, 2009. Beekeeping for Dummies [M]. 2nd ed. Wiley Publishing, Inc., Indianapolis, Indiana.

Cramp D, 2008. A practical manual of bee keeping [M]. Spring Hill House, United Kingdom.

Ruttner F, 1983 Queen rearing: biological basis and technical instruction [M]. Bucharest, Romania: Apimondia Publishing House.

Sammataro D, Avitable A, 1998. The Beekeeper's Hand Book [M]. 3rd ed. Cornell University Press, USA.

MODULE 10:
CARE FOR THE HEALTH OF HONEYBEE COLONY

>>> INTRODUCTION

Honey bees play a vital role in the environment by pollinating both wild flowers and many agricultural crops as they forage for nectar and pollen, in addition to producing honey and beeswax. The essential and valuable activities of bees depend upon beekeepers maintaining a healthy population of honey bees, because like other insects and livestock, honey bees are subject to many diseases and pests and toxic materials.

Successful beekeeping requires regular and on time monitoring of any factors that endangers honeybee life and threaten their products. Apart from identifying the occurrences and distributions of endangering factors, regular monitoring helps to think on devising prevention and/or control mechanisms.

This module is enables to identify and characterize the major honey bee diseases, pests and predators, develop efficient methods for prevention, control and management practices and follow the appropriate techniques to diagnose diseases and treatment procedures to maintain healthy and productive colony.

1 IDENTIFY AND CHARACTERIZE COMMON HONEYBEE DISEASES AND PARASITES

Honey bee (*Apis mellifera*) is prone to be infected with fungal, bacterial and protozoan pathogenic organisms. Honey bee diseases in Ethiopia include chalkbrood diseases caused by pathogenic fungi *Ascosphaera apis*, Nosematosis caused by *Nosema apis* and amoeba caused by a single protozoa *Malpighamoeba mellificae*. Many invertebrate pests belong to insects themselves such as ants, beetle, moths, lice, termites, mites, and large vertebrate animals such as amphibians, reptile, lizards, birds, mammals like honey badgers and mice were recognized as pests in the Ethiopian honey bees. Honey bee diseases, predators and pests are problems for bee keeping practice in Ethiopian (Desalegn, 2015).

The success of apiculture is influenced by these diseases causing pathogenic organisms and various pest animals. Infections of the disease ranging from chronic to highly virulent can result loss of honey bees' population and loss of honey bee products such as honey, wax and caused honey bees to abscond and death. The economic loss associated with the presence of honey bee diseases and pest was estimated in some works and significant loss was reported (Segeren, 2004; Cramp, 2008; Amssalu et al., 2010).

1.1 Bacterial Diseases

Bacteria are microscopic, primitive single-celled organisms, most of which are harmless. Some are beneficial. However, some are pathogenic to animals. The best known bacterial infections of bees are American foulbrood and European foulbrood diseases.

1.1.1 American Foulbrood Disease

American foulbrood (AFB) is probably the most serious of the brood diseases (Fig. 29). It is highly infectious bacterial disease and can be spread by drifting bees, by robbing and by the beekeeper moving from an infected hive to others during inspections. It is widely distributed wherever colonies of *Apis mellifera* are kept (MAAREC, 2005).

Cause: Microscopic spore-forming bacteria called *Paenibacillus larvae*. *Paenibacillus larvae* are a slender rod with slightly rounded ends and a tendency to grow in chains. The rod varies greatly in length, from 2.5 to 5 micrometers (μm); it is about 0.5μm wide. The spore is oval and about twice if wide, 0.6 - 1.3μm (Shimanuki and Knox, 2000).

Effect: American foulbrood is the most widespread and the most destructive of the brood diseases. At first, the strength of an infected colony is not noticeably decreased and only a few dead larvae or pupae may be present. The disease may not develop to the critical stage when it seriously weakens and finally kills the colony until the following year, or it may advance rapidly and seriously weaken or kill the colony the first season.

Transmission:

- ➢ The spores are fed to young larvae by the nurse bees. They then germinate in the gut of the larva and multiply rapidly, causing the larva to die soon after it has been sealed in its cell. By the time of death of the larva, the new spores have formed. When the house bees clean out the cell containing the dead larva, these spores are distributed throughout the hive and more and more larvae become infected.
- ➢ The honey in an infected colony becomes contaminated with spores and can be a source of infection for any bee that gains access to it. For example, as a colony becomes weak, it cannot defend itself from attacks by robber bees from strong nearby colonies; these robbers take back the contaminated honey to their own colony and start again the cycle of infection.
- ➢ The beekeeper also may inadvertently spread the disease by exposing contaminated honey to other bees or by interchange of infected equipment. Using stored equipment contaminated by spores will lead to a new infection even after years of storage.
- ➢ Drifting bees or swarms issuing from an infected colony may spread the disease

(MAAREC, 2005).

Symptoms:

> - *Stage/age of brood infected*: Attacks young larva (<2 days of age); kills in late larval to pupal stage
> - *Appearance of brood comb*: Irregular pattern of sealed brood with discoloured, sunken or punctured sealed brood
> - *Color of dead brood*: Dull white, becoming light brown, coffee brown, brown to dark brown or dark brown, or almost black
> - *Consistency of dead brood*: Soft, becoming sticky to ropy
> - *Odor of dead brood*: Slight to pronounced glue odor to glue-pot odor
> - *Character of scale*: Uniformly lies flat and adhere on lower side of cell; threadlike tongue of dead pupa adheres to roof of cell (Shimanuki and Knox, 2000).

1.1.2 European Foulbrood Disease

European foulbrood disease (EFB) is generally considered less virulent than AFB; although greater losses in commercial colonies have been recorded in some areas resulting from EFB (MAAREC, 2005; Fell, 2008).

Cause: microscopic lancet-shaped bacteria called *Streptococcus pluton*. It is an oval rather pointed bacterium, which grows entirely within the gut cavity of brood and it does not form spore (Coffey, 2007).

Effect: European foulbrood (EFB) is most common in the honey flow period when brood rearing is at its height, though usually the earliest reared brood is not affected. Sometimes the disease appears suddenly and spreads rapidly within infected colonies; at other times it spreads slowly and does little damage. European foulbrood may severely weaken a colony but usually does not kill an entire colony. As a rule, it subsides during honey flow period, but occasionally it continues to be active during this period. A good honey flow seems to hasten recovery.

Transmission:

> - The organism becomes mixed with the brood food fed to the young larva by the nurse bees, multiplies rapidly within the gut of the larva, and causes death within about 4 days after egg hatch.
> - House bees cleaning out the dead larvae from the cells distribute the organism throughout the hive.
> - Since the honey of infected colonies and the beekeeper's equipment are undoubtedly contaminated, subsequent spread of the disease is accomplished by robber bees, exposure of contaminated honey by the beekeeper, interchange of contaminated equipment among colonies, and perhaps to some extent, by drifting bees. The bacteria are not a spore forming species.

Symptoms: Larvae diseased by European foulbrood move restlessly within their cells and, therefore, when they die, are usually twisted in the cells (Fig. 30) (Shimanuki and Knox, 2000).

- *Stage/age of brood infected*: Usually young unsealed larvae; occasionally older sealed larvae
- *Appearance of brood comb*: Unsealed brood; some sealed brood in advanced case with discoloured, sunken or punctured capping
- *Color of dead brood*: Dull white, becoming yellowish white to brown, dark brown or almost black
- *Consistency of dead brood*: Watery to pasty; rarely sticky or ropy
- *Odor of dead brood*: Slightly to penetratingly sour in odour
- *Character of scale*: Usually twisted in cell; does not adhere tightly to cell wall and it is rubbery

1.2 Viral Diseases

A virus cannot live and multiply on its own and therefore relies on living body cells for survival. The resulting disease produced is due to the destruction or multiplication of cells which the virus has invaded. A virus is not affected by antibiotics, and so diseases caused by viruses are not effectively treated by administration of antibiotics. Sac-brood and bee paralysis are well-known viral disease affecting bees.

1.2.1 Sac-brood Disease

Sac-brood disease is perhaps the most common viral disease of honey bees. Several reports indicate that nurse bees are the vectors of the disease. Larvae are infected via brood-food gland secretions of worker bees (Ritter and Akratanakul, 2006).

Cause: Sac-brood disease is caused by virus called *Morator aetotulas*. Sac-brood virus particles are 28nm in diameter, non-enveloped, round and featureless in appearance (Coffey, 2007).

Effect: Sac-brood is a widely distributed disease, but it usually does not cause serious loss. However, the beekeeper should learn to recognize sac-brood, so it will not be mistaken for the serious foulbrood diseases. Sac-brood may appear at any time during the brood-rearing season, but it is most common during the first half of the season. Usually it subsides after the main honey flow starts.

Transmission: The virus is probably fed to the young larva by the nurse bees in the brood food. It multiplies rapidly within the larva until it causes death. Then the house bees cleaning out the cells probably distribute the virus to other larvae within the hive. The disease is usually limited to one or a few colonies in an apiary.

Symptoms (Shimanuki and Knox, 2000):

- *Stage/age of brood infected*: Usually older sealed larvae; occasionally young unsealed larvae (Fig. 31)

- *Appearance of brood comb*: Scattered cells with punctured sealed brood, often with two holes
- *Color of dead brood*: Yellow-grayish or straw colored becoming brown, grayish black, or black
- *Consistency of dead brood*: Watery and granular tough larval skin forms a sac.
- *Odor of dead brood*: None to slightly sour
- *Character of scale*: Head prominently curled up; does not adhere tightly to cell wall; lies flat on lower side of cell; rough texture; brittle

1.2.2 Virus Paralysis Disease

Two different viruses - chronic bee paralysis virus (CBPV) and acute bee paralysis virus (ABPV), have been isolated from paralytic bees. Honey bees infected with viruses generally fail to fly, appear lethargic and often crawl on the ground. Bees often have bloated abdomens and discoloured deformed wings. Infected colonies may collapse (MAAREC, 2005).

Symptom:

- The bees will be crawling around the hive's alighting board or entrance in a semi-moribund or moribund state, often in large numbers. These bees will not react if you prod them. And these bees are usually unable to fly.
- They often appear blacker in colour than other bees and shiny as they become hairless.
- A close examination will often show that these bees have extended abdomens.
- The bees are often refused entry to the hive. This situation looks remarkably like pesticide poisoning when bees are refused entry and die outside the hive in large numbers. Do not confuse the two. This situation can also be symptomatic of starvation, so this is yet another confusing signal (Coffey, 2007).

1.2.3 Deformed Wing Virus

Bee infected with deformed wing virus (DWV) disease generally emerge with deformed or poorly developed wings, bloated abdomens and discoloration, which is attributed solely to the feeding activities of Varroa on the developing larvae (Lea, 2015).

Symptom:

- DWV causes clinical symptoms in developing pupae, including pupal death.
- Newly emerged bees from affected colonies show deformed or poorly developed wings.
- The appearance/extent of deformity depends upon the stage at which individual bees become infected.
- Additional symptoms include a bloated, shortened abdomen.
- The virus multiplies slowly, and pupae infected at the 'white-eyed' stage of

development survive to emergence but are malformed and soon die.
➢ Brood may die earlier in development, and bees infected as adults appear normal until death. Colonies with DWV infecting adult bees or brood are likely to be half the size of virus free colonies.

1.2.4 Black Queen Cell Virus

Black queen cell virus (BQCV) is a viral disease. It infects developing queen larvae and causes them to turn black and die. It is thought to be associated with *Nosema*.

1.3 Protozoal Diseases

Protozoa are much larger than bacteria but still comprise a single cell. Many of the protozoa that cause disease require two hosts to complete the cycle. The most common protozoal diseases of honey bees are Amoeba and Nosema.

1.3.1 Amoeba

Amoeba is diseases of honey bee caused by a single celled parasite called *Malpighamoeba mellificae*. The parasite affects Malpighian tubules of honey bees and shortens the life span of bees. The disease was reported with high prevalence rate in different regional state of Ethiopia such as Oromia region with prevalence rate (88%), Amhara region (95%) and Benishangul-Gumuz (60%).

Study on annual cycle and seasonal dynamics of amoeba from the Holeta research center reported that amoeba cysts were highest cyst number (disease intensity) in the months of April and August and lowest intensity in the month of January. This study helps to understanding the seasonal dynamics of the diseases in the area and to undertake seasonal management of colony honey bees (Desalegn and Amssalu, 1998).

Symptoms: Characteristic symptom of amoeba is dysentery.

1.3.2 Nosema Disease / Nosemosis

Nosema disease is generally regarded as one of the most destructive diseases of adult bees, affecting workers, queens and drones alike that affects colony development, queen performance and honey production.

Cause: Nosema (Nosematosis) is caused by a small single-celled protozoan called *Nosema apis*. Nosema spore are 5 to 7mm in length. It infects the intestinal tract of adult bees.

Distribution: In Ethiopia, Nosema was also reported from different regions with varying prevalence rate such as 58% in Oromia, 60% in Benishangul-Gumuz and 47% in Amhara regions. In the central highlands of Ethiopia, highest infestation level of *Nosema apis* and spore number per individual honey bees was reported in the month of August and September. The study also found positive correlation between Nosema infestation rate, number of *Nosema* spore per individual honey bee and humidity (Amssalu and Desalegn, 2005).

Effects:

> The spores of *Nosema apis* enter the body of the adult bee through the mouth and germinate in the gut. After germination, the active phase of the organism enters the digestive cells that line the mid-gut of the adult bee where they multiply rapidly.
> The contents of these cells are used as a food supply until reproduction ceases and new spores are formed.
> The cell then ruptures and sheds the new spores into the mid-gut where they pass down through the small intestine to the rectum. Here these accumulate and are voided in the excreta of the bee.
> The cycle begins over again when the spores contaminate the water or food of other bees. Spores will remain viable for many months in dried spots of excreta on brood combs. They lose their viability within a few days in water exposed to direct sunlight, and they are also easily killed by heat and by some fumigants.
> In climates with pronounced long periods of flight restrictions, i.e. no flight opportunities even for a day, the infection easily reaches a severe stage that visibly affects the strength of the colony. Less obvious infection levels in other climates often go undetected. The damage caused by Nosema disease should not be judged by its effect on individual colonies alone as collectively it can cause great losses in apiary productivity (Amssalu and Desalegn, 2005).

Transmission: The spread of Nosema disease occurs chiefly because of the use of contaminated equipment, infected package bees, infected queens and her attendant workers, and the robbing of infected hives.

Symptoms:

> Mid-gut of infected bees will be discoloured.
> Seriously affected worker bees are unable to fly and may crawl about at the hive entrance or stand trembling on top of the frames.
> Damage to the digestive tract may produce symptoms of dysentery. Infected workers, unlike healthy ones, defecate in or on the outside of the hive rather than in the field. Dark brown coloured faeces (excreta) on the top bars and at hive entrance.
> The bees appear to age physiologically: Their life-span is much shortened, and their hypo-pharyngeal glands deteriorate, the result is a rapid dwindling of colony strength.
> A pile of dead bees on the ground in front of the hive may be manifestations of *Nosema* infection, but they may also be caused by other abnormal conditions.
> However, if crawlers or unusual numbers of dead bees are seen in the apiary or if a colony fails to build up properly, Nosema disease should be suspected.

> To ascertain whether Nosema disease is present requires microscopic examination of the abdomens of older adult bees.

1.4 Fungal Diseases

Fungi (or moulds) vary both in size and cell structure. They have a low pathogenicity; that is, their ability to produce disease is limited. Chalk-brood and Stone-brood are common fungal diseases.

1.4.1 Chalkbrood Disease /Ascosphaerosis

Chalkbrood is an infectious disease of honeybee larvae caused by a fungus *Ascosphaera apis*, which causes death and mummification of sealed brood of honeybee with consequent weakness of the colony (Fig. 32).

Distribution: In Ethiopia the geographical distribution of chalkbrood diseases in honey bee were recorded. The study reported an infection rate of 37.12%, 19.89%, 17.93% and distribution rate of 87.50%, 56.56% and 33.33% in Amhara, Oromia and Benshangul-gumuz. The finding shows that moist Dega, moist Weina Dega and wet Weyna Dega were identified as suitable ecological zones. However, the dry Alpine, dry Bereha and moist Bereha areas are not suitable for the diseases at all. The effect on productivity indicated that the mean yield of honey in colony infected with chalkbrood diseases (45kg) is lower than the mean yield (80kg) in uninfected bee colony (Desalegn, 2000; Aster et al., 2010).

Symptoms: Initially, the dead larvae swell to the size of the cell and are covered with the whitish mycelia of the fungus. Subsequently, the dead larvae mummify, harden, shrink and appear chalklike as shown in Fig. 32 (Shimanuki and Knox, 2000).

> *Stage/age of brood infected*: Usually older sealed larvae, up right in cell
> *Appearance of brood comb*: Sealed or unsealed brood
> *Color of dead brood*: Chalk white; sometimes mottled with black spots
> *Consistency of dead brood*: Larva can be removed from cell as a fluid-filled sac, watery paste like.
> *Odor of dead brood*: Slight non-objectionable
> *Character of scale*: Larval 'skin' remains intact; does no adhere to cell wall; brittle, chalky white, mottled or black

1.4.2 Stone-brood Disease

Stone-brood is a fungal disease primarily caused by *Aspergillus flavus* and to a lesser extent *Aspergillus fumigatus*. Both larvae and pupae are susceptible and the disease causes mummification of developing larvae.

The 'mummies' are covered with a powdery green growth of fungal spores, with the highest concentration of fungal spores near the head of infected larvae and pupae. The diseased larvae are solid 'mummies' and not sponge-like as in chalk-brood.

This disease is only considered of minor importance and no treatment is necessary. House-bees normally remove infected brood and the colony recovers naturally. It should be noted that *Aspergillus* moulds can cause respiratory problems in humans and thus it is important not to sniff or inhale infected combs (Ritter and Akratanakul, 2006; Coffey, 2007).

1.5 Parasites

A parasite is an animal or plant which lives on or in an organism of another species (the host), from the body of which it obtains nutriment. Parasites may be internal (within the body) or external (on the body) of the host animal. Parasites may affect different stages of the life cycle of the host animal (or bee). Parasites can cause disease in many ways. They can either directly damage the host animal or they can introduce infectious organisms into or onto the host. In these situations, the parasite is called a vector of disease.

If the parasites accumulate in large numbers in the host, they cause the animal to lose condition by competing directly, reducing the surface area, or by damaging the intestine wall to reduce the absorption of nutrients. Parasitic diseases are particularly severe in young animals because they are more susceptible. They retard growth and lower resistance, thereby making the animal also more susceptible to other problems, including environmental stress, poor nutrition and infections. The tracheal mite, Varroa mite and bee louse are the most common honeybee parasites.

1.5.1 Varroa Mite / Varroatosis

Varroa mite (*Varroa destructor*) is the most serious pest mainly threatening beekeeping industry development all over the globe by causing huge honeybee colony mortalities.

On top of the ability to replicate only in a honeybee colony, *Varroa destructor* is an external parasitic mite that can be attached on (to) the external body of adult bees and capped brood and suck hemolymph (blood) that results in a disease called varroatosis. Through repeated feeding, the mite shortens the life span of the honeybees, weakens the honey bee colonies and thus decrease honey production and causes the ultimate perishing of colonies (Desalegn, 2014).

Cause: Only mature female mites survive on adult honey bees and can be found on both workers and drones and rarely on queens (Fig. 33). Varroa mites are reddish brown in color, about the size of a pin head, and can be seen with the naked eye (1.6mm wide × 1.1mm long). Their flat shape allows them to squeeze between overlapping segments of a bee's abdomen to feed and escape removal by grooming bees. Their flat shape also permits them to move easily in the cells of developing bee brood. Male mites are smaller and light tan in color. Adult males do not feed and are not found outside of brood cells. Adult females of *V. destructor* are found inside brood cells or walking rapidly on comb surfaces (Coffey, 2007).

Symptom:

➢ Individual mites are often seen clinging tightly to the body of adult bees, mostly on

the abdomen, where the segments overlap, between the thorax and the abdomen and at the ventral entry.
- Colonies destroyed by the Varroa mite are often left with only a handful of bees and the queen, the other bees having died during foraging or having drifted to neighbouring colonies, where the mite population can increase before killing these colonies also. In this way mites may cause colonies to die.
- The presence of adult bees crawling on comb surfaces or near the hive entrance usually indicates a late stage of heavy mite infestation.
- In bees which have been more severely infested, deformed wings, bloated abdomens and discoloration of legs are typical symptoms of infection.

1.5.2 Tracheal Mite /Acaropisosis

Tracheal mite infection (known as acarapisosis or acarine) is an infestation of the respiratory system of adult bees by the parasitic tracheal mite *Acarapis woodi* (Lea, 2015). It infects worker, drone and queen honey bees, and can be serious if not treated (Nasr, 2014). This internal parasitic mite lives within the tracheae, or breathing tubes, inside the thorax of adult honey bees.

Mites usually infest adult bees when they are less than 3 days old, but older (up to 10 days) bees are also susceptible, particularly within the winter cluster. The mite reproduces inside the trachea leading into the thorax from the first pair of spiracles. A honey bee becomes infested when a female mite crawls through the spiracles and enters these tubes, attracted by the vibration of the wing roots, and by puffs of air coming out during respiration.

The mites pierce the breathing tube walls with their mouth parts and feed on the hemolymph, or blood, of the bees. She quickly lays eggs inside the tracheae, usually no later than one or two days after initial infestation (MAAREC, 2002).

Symptoms:

- Adult bees that are infested with tracheal mites will cluster in front of the hive, appearing confused and disorientated, unable to return to the colony.
- Large numbers of bees may also be seen crawling up stems of grass in front of the hive.
- Such behaviors are not, however, clear indications of acarapisosis: not only are they associated with other pests and diseases; even bees that are severely infested with mites can behave in a normal way, although their tracheal wall has been damaged.
- Detection under a low power microscope after simple dissection of the bees is the only reliable method of diagnosis (Lea, 2015).

1.5.3 Bee Lice

Bee lice (*Braula coeca*) are wingless ectoparasite fly which causes significant damage in colony bees. Bee lice larvae feed on honey and pollen by tunnelling under the cell capping.

The adults are small (slightly smaller than the head of a straight pin), and reddish-brown in color. Although several adult flies may live on a queen, usually only one will be found on a worker.

The adult lice feed on nectar directly from the mouth of honey bees, this reduce food availability of queen and reduce egg-lying capacity. In Ethiopian infestation of lice in honey bees was reported in different regions.

An experimental study had shown bee lice to be an evident cause of reduction in the number of worker bees and honey production where treated colonies with tobacco smoking against bee lice had produced an average of 18kg of more honey as compared to the untreated ones (Adeday et al., 2012).

2 PESTS AND PREDATORS

2.1 Pests of Honey Bee

Practical knowledge on the identifications of honeybee pests that endangers the life and products of honeybees and developing appropriate control measures is largely a question of success in the beekeeping sector. With this understanding, the existing literature indicated that many assessments were conducted in different parts of Ethiopia at different times with the objectives of identifying local honeybee pests along with their distribution ranges and kinds of products they affect. Accordingly, the review of the findings indicates as more than 15 honeybee pests identified and recorded in the country with the products types they are affecting. According to these studies, ant (different types), wax moth (greater and lesser wax moths), mice, birds (different types), honey badger, wasps, death's head hawks moth, beetles (different types), lizards, toads/frog, prey-mantis, spiders, pseudo scorpions (*Chelifer* spp.) were among the major honeybee pests registered locally (Desalegn, 2001; Adeday et al., 2012).

2.1.1 Wax Moth

The greater wax moth, *Galleria mellonella* (Fig. 34) is the most serious pest of honeycombs. They are silvery-grey/brown to dull in color and 1.3 – 1.9cm long.

These moths are an especially serious problem in tropical and subtropical climates, where warm temperatures favor their rapid development. Female greater wax moths lay their eggs in a cluster, usually in the cracks or between the wooden parts of the hive. The larvae are the destructive stage. They obtain nutrients from honey, castoff pupal skins, pollen, and other impurities found in beeswax, but not from the beeswax itself. Consequently, older combs are more likely to be damaged than new combs or foundation (Shimanuki and Knox, 2000).

Wax moth usually found in most beehives, and the bees normally repair any damage it causes as soon as it appears. It is for this reason that most beekeepers are unaware of it until the bees lose their ability to defend themselves and the moth larvae take over.

Effects: The larva move through the comb, eating honey, pollen and beeswax. The tunnels they make through the comb are silk lined and full of frass. These tunnels are easily seen and are just the beginning. When they have grown (up to about 3cm for the greater wax moth), the grubs will hollow out a shallow, boat-shaped depression in the woodwork, spin a cocoon and pupate. They do this in large numbers and, if you are unfortunate enough to see this, you will know you have left things very late.

Symptoms:

- Grey caterpillars 0.5 to 2.5cm long can be seen scrawling over the top bars when you remove the inner lid.
- Combs are affected and the spaces between the combs are covered in spider webs.
- Whole combs are eaten up and changed into a grey-black mass spun together with webs.

Preventive measures:

- The only protection against wax moth is to keep your colonies strong and healthy.
- If a colony is failing, consider uniting it with another one (after checking it has no disease). Protecting stored comb is difficult. Wax moths won't usually infest clean comb that has no pollen or other debris in it, and they never attack foundation. They will attack, for example, comb with honey in it, or comb containing pollen, brood, old brood remains, and cocoons and so on.
- Freezing comb kills all stages of the pest. When you store comb keep the boxes in a cool, well-ventilated place with a spacer between them to let in light: wax moths shy away from light.
- In cold climates you can store your supers on top of your hives with a mat or escape board between them. This will allow limited bee access but will keep the supers cold.
- Always remove frames from the colony which cannot be occupied by bees and store these in a protected place.
- Transfer lightly affected frames to the strong colonies. The bees will clean them and repair them.
- Put the affected frames in a closed area and treat them with glacial acetic acid or smoke of burning sulphur.
- Store frames with good comb in an airtight container or box which you disinfect from time to time as mentioned above. Watch out for mould.
- Ensure that the colonies are strong and have adequate food stores.
- Adapt the hive space to the strength of the colony.
- Reduce the hive entrance.
- Seal cracks and crevices in hive walls.

- Protect the colonies against pesticide poisoning.
- Control pests and diseases that might otherwise weaken them.
- Remove any wax and debris accumulated on the bottom boards of the hives (Segeren, 2004).

Treatment measures: The control treatment measures are fumigation and heat treatment (Ritter and Akratanakul, 2006).

- *Fumigation* is the usual treatment; new combs should be treated less frequently. Among the most commonly used fumigants are naphthalene, ethylene dibromide and methyl bromide. All, including paradichlorobenzene, are very poisonous to bees and humans and, in addition, lead to residues in honey.
- *Heat treatment of wax*: The development of wax moths can be interrupted for several months if the combs are heated at 48℃ for three hours. All treatments should be repeated at intervals depending on the level of infestation. Regular control is therefore recommended.

2.1.2 Ants and Termites

Among so many pests existing in Ethiopia, ant (*Dorylus fulvus*) is found to be the most troublesome to honeybees and beekeeping sector. Ant causes great damage to honeybees and their products. As a result, many productive bee colonies have been either killed or abscond due to intolerable ant attack (Desalegn, 2007).

Ants (*Dorylus fulvus*): Brown or black insects, 0.5 to 2cm long, which operate in groups and attack weak colonies by consuming honey and dragging out brood. Some kinds of ants operate only at night (Segeren, 2004). Ants are most troublesome to honey bees and bee keeping sector. Ant eats or carries off any comb contents honey, pollen and brood. Ants are one of important honey bees' enemies and causing a serious problem.

Ants causes severe economic loss in honey production by killing bees, rob their products, initiate aggressiveness in bees, lead to absconding and destroying the entire colony of honey bees. Bees are the first and most victim of the attack with ants followed by honey. Ants were also reported to feed on bees wax and pollen after all honey and the broods are depleted. Study in seven districts of West Shoa Zone, shows that 44% of the colony bees were attacked by ants result in 24% of absconding and 4.2% death and 29% of honey production was loss to due to ant's attack. The total economic loss caused due ants attack was estimated to be 3,839,810 Ethiopian birr. Comparing the monetary loss in terms of products, the loss through bee products (honey and wax) overweight the loss produced through the bee colonies themselves (Desalegn, 2007).

Symptoms:

- Extremely restless bees and a buzzing sound near the flight entrance
- Dead and half eaten bees around the hives
- The hive has been abandoned by the bees

- Wood with holes (only soft wood will be affected) (Segeren, 2004)

Preventive measures:

- Provide a good barrier between the ground and the hives (Fig. 10.1). You can do this by suspending the hives with wire on air to prevent climbing of ants to the hive (Fig. 10.1 A), placing the legs in a shallow pan of water/used car oil (Fig. 10.1 B), or placing an apron or shield with cone shaped metal cup on hive stand (Fig. 10.1 C).
- Spread a combination of ash, grease, and ground cinnamon and/or garlic powder circumferentially around the legs.
- Make sure that no weeds grow below the hives, allowing the ants to climb up to the hives.
- Make the bee hives and stands of fairly hard wood known to be resistant to termites.
- Treat the wood with creosote (Amssalu and Desalegn, 1999).

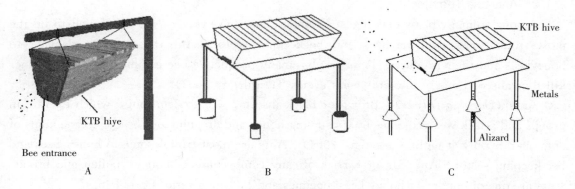

Fig. 10.1 Ants attack protection methods
A. Hanging the hive   B. Stand legs in water/oil tins   C. Cone shaped metal cup on hive stand

2.1.3 Small Hive Beetle

Small hive beetle (SHB), *Aethina tumida* is indigenous to Africa, where it is considered a minor pest of honey bees. Both adults and larvae can be serious pests of weakened honey bee colonies or honey supers. The beetles multiply to huge numbers, their larvae tunnel through comb to eat brood, ruin stored honey, and ultimately destroy infested colonies or cause them to abscond. The beetle also defecates in the honey, causing it to ferment and run out of the combs. In Ethiopia the small hive beetle was recognized as local honey bee parasite in two honey flow seasons in most part of the regions (Amssalu and Desalegn, 2008).

The newly-matured adult beetle is light, yellowish brown and becomes brown, dark brown and finally black at full maturity. The beetles are ovoid in shape and approximately 5 - 7mm in length. Their antennae are club-shaped and have a short wing case (elytra), which is covered with fine hairs (Coffey, 2007).

Effects: SHBs are typically opportunistic predators that don't cause the demise of strong colonies; they are more problematic in the Deep South in areas of sandy soil. SHB females lay eggs en mass on or near pollen (bee bread). Eggs can hatch within one day. The larvae (5 - 14 days) damage comb while feeding on pollen and damage honey by carrying yeast that causes its fermentation. The yeast is very repellant to bees and may lead to absconding. Larvae then travel on the ground and pupate in the soil. SHBs are attracted to weak, stressed bee colonies and pollen in stored comb. Treatment thresholds have not been established, but fewer than 100 adult beetles per hive (that have not begun reproduction) are probably safe. Hygienic bees are good at finding egg masses and removing them (Amssalu and Desalegn, 2008).

Detection: A simple technique used to look for beetles is to remove the lid and place it upside down on the ground. Place the brood box on top of the upturned lid. If beetles are present, they will move out of the brood box away from the light and may be seen crawling in the lid. Beetles may also be seen 'surfing' across the comb if the frame is exposed to sunlight (Coffey, 2007).

Control and preventive measures:

- Regular hive inspections
- Maintenance of strong, queen right colonies
- Removal of all equipment (brood chambers and frames) housing any dead colonies or 'dead outs' from the apiary in a timely manner
- Avoid providing more chambers than the colony can patrol.
- Avoid discarding burr comb in the apiary (collect it instead).
- Avoid providing more pollen than the colony can consume within 5 days during supplemental feeding.
- Minimal storage time of honey supers (<3 days) prior to extraction
- The use of a low humidity environment for any honey super storage prior to extraction
- Timely processing of wax
- Avoid storage of left over products of extraction ('slum gum').
- Allow bees to clean and dry out wet extracted honey supers from their own hives.
- Freeze infested frames to kill SHB larvae and eggs and if the damaged area is small, remove the SHB nest, wash the frame vigorously with sprayed water, and return it to a strong colony to repair.
- Discard moderate and heavily damaged comb.
- Preparation of in-hive beetle traps (such as Beetle Jail, AJ beetle eater, Cutt's beetle blaster), containing food-grade mineral oil, organic vegetable oil or apple cider vinegar (Amssalu and Desalegn, 2008)

2.1.4 Other Insects

Other insects such as fire ants, hornets, and yellow jackets are typically opportunistic predators that cause little damage to strong colonies. Remove chambers on the hive than what the bee colony can patrol.

2.2 Predators of Honey Bees and Their Products

Bee keeping in tropical climates frequently suffers from damage caused by different predators (Fig. 35). In Ethiopia honey bee predators were reported from all the class of animals including: reptiles, man, cattle, predatory birds, skunk, raccoon, monkey or apes, mice and honey badger (Segeren, 2004; Haileyesus, 2015).

2.2.1 Predatory Birds

Bees are virtually defenceless against predation of by birds. The heavy traffic of bee flying in out of hive provides as source of food for different bird species. For example, honey guide birds/bee-eaters (*Merops apiaster*, *Merops orientalis*), swifts (*Cypselus* spp., *Apus* spp.), drongos (*Dicurus* spp.), shrikes (*Lanius* spp.) and wood peckers (*Picus* spp.) are reported as predators of honey bees (Caron, 2000).

In Ethiopia bee eating birds are described as problem in colony of honey bees. In Illubabor zone birds locally known Hamaa were reported to eat bees by breaking the hives at night and other bird called Simbiroo predate bees by waiting around the hive. These birds weaken the hive and reduce the quantity of honey harvested (Haileyesus, 2015).

2.2.2 Man

Perhaps the greatest predator/pest of honey bees is man. Bees can be vandalized, stolen and/or burned. In the southern regions of the country, hunting of wild honey bees is practiced. Peoples never keep bees, but harvest wild bees from caves, cracks using indicator birds and numbers of bees were burned through fires. Man is also reported as predators or pests and robbery of bee were reported in rural bee keeping areas of Ethiopia and developing countries. Some authorities speculate that bees came to Ethiopia from Egypt along the Nile Valley, and that the same bees were also taken to Somalia. Somali bee-eater is one of the most serious pests of honeybee colonies in Somalia. In addition, reduction of bee forage by large-scale agriculture and urbanization is deleterious to bee populations.

2.2.3 Mice

Mice can destroy comb and may inhabit honey houses (Caron, 2000).
- Use the smallest entrance possible to prevent entry into the hive. Chase them away, replace any damaged comb and frames, and wash any urine from the interior surfaces of woodenware with water only.
- Mouse traps or non-synthetic mouse repellents in honey houses
- Mouse poisons in honey houses, mouse traps or repellents in bee hives

2.2.4 Honey Badger

Honey badger destroys hives in search of honey, brood and adult bees, which they eat.

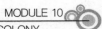

A recommended preventive measure includes:
- Selection of apiary sites away from known badger/bear habitat, or if in a known habitat, away from streams and ridges.
- Place at-risk apiaries near a dog.
- Use of two straps per hive to deter bears/badger.
- The installation of an electric bear fence
- Critter getter type alarms
- Shooting bears
- Poisoning bears

2.2.5 Reptiles (Snake, Toads and Lizards)

Reptiles including snake, toads/frogs and lizards are also a problem for colonies in areas of the tropics. They can eat large numbers of bees from the entrance of the colony at night or capture bees as they leave the hive early in the morning. If toads are eating bees from the colony, toad feces which contain bees will be seen around the front of the hive.

In areas where ants and toads are a problem, it is necessary to have the hives on hive stands to prevent losses. Hive stands make working the hives easier and can also lessen termite damage to wooden hives. Using termite resistant posts to make the stands can prevent termites from gaining access to the hives.

Harassment from animals

When animals play games they may knock hives, brood that has been pulled out of the frames, hoof or paw-prints. They can cause considerable damage and suffer considerable damage themselves when they come too close to or get into the hives.

Prevention: If you experience disturbances of this kind, fence in the apiary with a fence of poles, wire or bamboo or a hedge.

Robbing

In dearth periods when the bees detect honey or sugar solution in the hives of other colonies, they attempt to remove these. Strong colonies can defend themselves against this, but weak colonies will not be able to do so and are either killed in their attempt to defend the hive or abscond after a while.

Symptoms:
- Flights between bees at the flight entrance
- Dead bees under the hive
- The colonies are restless and irritable.

> *Prevention*:
> - Never spill honey or sugar solution outside the hives.
> - Feeding must always be done in the hive so that the bees only have access to the sugar from inside the hive. It is best to give the sugar solution in the evening.
> - In weak colonies, always make the flight entrance small (5cm).
> - Ensure that the hives are well closed and have only one flight entrance.

3 PESTICIDE POISONING

Death of an insect by poisoning is generally caused by the failure of the alimentary system or the poison affects the nervous system leading to complete lack of coordination of the normal bodily functions, which in both cases leads to death and starvation (Coffey, 2007).

Symptom: The most apparent indication of serious poisoning is the sudden loss of adult bees. This loss is characterized by the appearance of many dead or dying adult bees and sometimes pupae at the hive entrance. However, in many instances the bees are lost in the field before returning to the colony. If only the foraging population is affected, the colony will start to recover in approximately 2 weeks, as new brood hatches out and house-bees become guards and foragers by natural progression. If, however, the house-bees are also poisoned by feeding on contaminated honey and pollen, the colony will be reduced to the queen, which is generally unaffected due to her royal jelly diet, and a small number of bees. Under such circumstances the brood will exhibit symptoms of neglect and poisoning, and bees often abscond.

Types of pesticides: In general, pesticides are classified into five main groups namely organophosphates, chlorinated hydrocarbons, carbamates, dinitrophenyls and botanicals. The symptoms of poisoning exhibited by honeybees may be an indication of the class of pesticide involved (Coffey, 2007).

- *Organophosphorous* poisoning causes bees to regurgitate, become disorientated and lethargic. A high percentage of bees die at the colony.
- *Chlorinated hydrocarbons and carbamates*: Bees affected by chlorinated hydrocarbons and carbamates display erratic movements followed by paralysis. In the former, many bees die in the field and at the colony, while in the latter, the colony becomes aggressive and most bees die at the colony.
- *Dinitrophenyls and botanicals* show a combination of both the above symptoms, bees regurgitate, giving them a wet appearance, followed by a display of erratic movements and finally paralysis. Infected bees die both at the colony and away from

home.

Prevention measure: The main source of poisoning is from agricultural sprays. Bees are caught in sprays in three ways: 1) when the bees are foraging on the sprayed crop, 2) when the crop is sprayed for weeds which are been utilized by the bees, and 3) when bees are flying over a crop which is being sprayed to reach a crop further away.

Thus, application method and time of spray is critical to minimize damage done by spraying. Tractor mounted sprayers are less harmful than air craft and helicopter spraying. Also, if a crop needs to be sprayed, it should be carried out before 8:00 am or after 8:30 pm or when flying activity is minimal.

Beekeepers can minimize damage to colonies by collaborating with farmers within flying distance of the hives, partially or completely closing colonies and moving colonies more than 5km from the area to be sprayed. Closing colonies is fraught with danger, larger colonies as these tend to overheat, causing brood loss, suffocation and melting of wax. Moving colonies is also impractical especially for beekeepers with relatively large numbers of hives. Thus, collaborating with farmers in the area is often the only option to the beekeeper:

- Keep in touch with the farmers and make them aware of the detrimental effects. Together try to find a less toxic insecticide.
- Ask the farmers to let you know when they are going to spray.
- Provide the hives with a ventilation screen, sprinkle water over the screen and close the flight entrance. The hives must remain in the shade and be watered every day. Open the hives again after 2 days.
- Temporarily move the hives from the area.

4 PREVENT AND CONTROL HONEYBEE DISEASES AND PARASITES

The best preventive and control measures for honey bee diseases and parasites or pests is to develop what is known as an integrated pest management (IPM) system. This is the management of pests employing a combination of methods that include economic, ecological and toxicological factors while emphasizing biological (as opposed to chemical) controls and economic thresholds. The basic components of an IPM programme are:

- Prevention and awareness (by regular inspections and thorough knowledge)
- Observation and monitoring
- Intervention (where necessary)

4.1 General Measures

In Africa, it is virtually impossible to keep honey bee colonies free of diseases and parasites for long periods of time. Drifting, robbing and foraging on the same blossoms are among the most common means of disease transmission and parasite transfer. While each specific honey bee disease or parasite calls for its own specific control methods, the following general recommendations, if properly adopted, can assist in preventing or at least reducing damage to honey bee colonies (Ritter and Akratanakul, 2006).

> Strenuous efforts should be made to maintain vigorous colonies, with large, healthy worker populations, good laying queens and adequate honey and pollen stores. This is only possible with a constant sufficient pollen and nectar supply.
> The number of hive boxes and combs should be adapted to the colony strength.
> Diseases, parasites and predators likely to significantly weaken colonies should be properly controlled.
> Apiary sites should be selected with much care: strong winds, damp, unhygienic conditions and lack of food should be avoided.
> Colonies must be protected against poisoning by pesticides: frequent surveys should be made of the level and types of pesticides used within the foraging range of the bees.
> All hive parts and equipment should be kept clean and in good working order.
> Hives should be kept on stands, and apiaries should be securely fenced, whenever the danger of predators renders these precautions necessary.
> Hives should be manipulated with great care; all practices likely to induce robbing or cause bees to drift should be avoided, including overcrowded apiaries.
> Raising awareness of neighbours, farmers and others about the benefits of the bees for health products and pollination may create better agricultural practices and thus better foraging and less toxicity to bees. Awareness rising could therefore be considered a very effective preventive method as well as one that increases productivity.
> Hive disturbance, by beekeepers, outsiders and/or other non-beekeepers, should be kept to the absolute minimum.
> The utmost care in the choice and use of chemicals for disease control cannot be overemphasised, as most of these substances easily contaminate hive equipment and honey, create resistance in the pathogens and weaken the bees.
> Organic beekeeping methods describe and rely on control methods that are beneficial to the bees, bee products and human health.

4.2 Traditional Methods

Beekeepers practice different prevention methods but are not totally efficient which requires developing suitable prevention methods. Different types of pest and predators control methods are employed in different bee keeping areas of the countries. These numbers of traditional control methods are either used individually or in combination with other methods in all types of bee keeping practice and in all hive types. Some of the cultural practices are as follow (Haylegebriel, 2014):

> *Ants*: Applying ash under the hive stands, clean the underneath of the hives & keep

their apiary neat, use of leaves of eucalyptus & aje (local naming) as deterrents when it appears, wrapping the hive stands with polyline bag, hunting and killing ant queens and using of another small ant *Cremato gasterchiarinii* as biological control
- *Wax moth*: Clean apiary, remove old comb, and strengthen the colony, fumigation with cotton cloth and sorghum bran, rubbing with recommended plant materials like *Vernonia amygdalina*, spraying garlic juice
- *Bee lice*: Clean apiary, fumigate with *Olea africana* and cigarette and sorghum bran and make the colony strong
- *Beetles*: Clean apiary, narrowing the hive entrance, hand picking and kill, cover opening of hive
- *Lizard*: Clean apiary, use spin around and kill
- *Snake*: Clean apiary, smoking with plant material and kill
- *Birds*: Putting something (cloth, festal…) and spin around the hive and kill
- *Honey badger*: barriers putting like thorny woods around the tree; fixing smooth iron sheet on trunks of a tree where hives are hanged, hanging hives on *Ficus* trees which has very smooth bark which is not suitable for honey badgers to climb, fastening corrugated iron on the bark of the trees containing honey bee colonies

Several preventive or control management practices to minimize the effect honey bee pests are practiced in the country. For example, strengthening honey bee colonies via feeding, removing unoccupied suppers and combs, trapping adult wax moths were tested against wax moths and results in reduction of infestation level of wax moths by 82.3% and increasing honey bee. Three different ant protection methods such as inner tube, smooth iron sheet and tin filled with used engine oils are tests for their effectives in preventing access of ants by exposing to massive raids of ants and the tin filled with used engine oil methods was found the best in totally protecting honey bee colonies.

4.3 Recommendations

The following methods either individually or in combination with other integrated methods of diseases and pest control are recommended to improve productivity honey bees and their health welfare.

Good hive management

The effect of different honey bee diseases and pests can be reduced by improved management techniques such as strengthening colony with bees or hatching brood and enlarging colony entrance to aid ventilation. Good bee keeping practice such as avoiding use contaminated equipment, transfer of infected combs from infected hives were recommended to avoid horizontal and vertical transmission of different honey bee diseases from colony to another near colony. In general, the following colony management tools with integrated other improved beekeeping practices are recommended to maintain a strong bee in each hive:

- Inspect every hive at least once a month.
- Move the hive to disrupt the life cycle of honey bee pests. Maintain close mowing or bare ground around the hive to facilitate chemical controls and provide less shelter for beetle larvae leaving the hive to pupate population.

Searching veterinary medicinal plants

For centuries, medicinal plants were used for treatment of different diseases in animal and humans in health welfare.

Crude medicinal plant extracts such as *Allium sativum* Linn., *Eugenia caryophyllus* Bullock & Harrison, *Piper betle*, *Curcuma longa* Linn., *Illicium verum* Hook, *Cinnamomum cassia*, *Rhinacanthus nasutus* Kurz, *Azadirachta siamensis*, *Acorus calamus* Linn., and *Stemona tuberosa* Lour. show anti-fungal activity for treatment of chalk brood disease. To prevent and control honey bee's diseases, it is recommended to search and investigate veterinary medicinal products which value for treating honey bee diseases.

Breeding resistance variety bees

Some hives appear to be more resistant to different honey bee diseases than others due their hygienic behavior. Theses colonies have an ability of their adult bees to uncap and remove affected brood which reduce the spread of infection to whole colony. This hygienic behavior was explained as being controlled by two recessive genes, one for uncapping and one for removal of larva. So, it is recommended to selective breeding of varieties of honey bees colony by evaluating the hygienic behavior in different honey bee species found in the country. Breeding resistant variety of colonies should be practiced through selective rearing of queen from the resistant colony of bees.

Providing education to bee keepers

Controlling of honey bee disease and pests require intensive knowledge of recognizing and detection of these different diseases causing organisms and pests in colony of honey bees. It is recommended to provide educational delivery through extension service to bee keepers on management of honey bees, detection of diseases in brood and how to manage the problems.

Quarantine measures

The spread and entrance of some honey bee diseases is associated with delivery of contaminated apiary equipment's such as wax foundation sheets. Establishing quarantine measures through legal regulation and enforcement on introduction of honey bees and equipment can be reducing introduction of diseases from infested areas. During introduction of colony bee to a given locality of providing health of stock of bee and educating bee keepers the risk of buying infected colony of is recommended. In general, the following measures are recommended when buying, selling of stock honey bees:

- Before introducing new species or race of honey bee, it is important to study its potential (diseases and pest resistance) quality, foraging behavior and availability of forage.
- In terms of introducing and buying of stock of bee from market/regions to locality areas, well quarantine and inspection is essential.
- During introduction of honey bee equipment from abroad to country, proper inspection of instruments is recommended.

5 DIAGNOSE AND TREAT INFECTED HONEYBEE COLONY

Monitoring and inspection for bee diseases is an important part of beekeeping. Apiary beekeepers must be able to recognize bee diseases and parasites and follow the appropriate techniques used to diagnose diseases and treatment procedures to maintain healthy and productive colony. Before proceeding to disease diagnosis and treatment measures, it is important to understand the characteristics of healthy colony for comparison purposes.

5.1 Characteristics of Healthy Honeybee Colony

Observing the health and condition of your colonies is an essential part of beekeeping. During your routine work, you will have many opportunities to observe the health and condition of adult bees (Caron, 2000; Fell, 2008).

- Each time you visit the apiary, check any weak colonies or dead ones and find out why they died. However, remember that often the strongest hive in the apiary may be the first hive to pick up a brood disease. Strong hives have more foragers and therefore more robbers to rob honey that may carry disease spores.
- In addition, at least twice each year, you should examine all your colonies' brood by shaking the bees from the brood combs. The most suitable time to carry out brood inspections is on a clear sunny day.
- Identifying the signs of disease and pests requires good light and good eyesight. If you need glasses for reading, wear it inspecting brood and adult bees.

First steps

- Make sure you are wearing suitable protective clothing-if you suspect that you are looking at signs of an emergency pest or disease, then you need to have clothing and footwear (or coverings) that can be decontaminated (or burnt).
- Check tools and equipment required to open a hive, including equipment you will need to take samples for diagnosis of pests and diseases.
- Obtain a supply of water and detergent.
- Check for any OHS hazards and take steps to minimize risk to yourself and others.

Hygiene between hives and apiaries

- Always wash your gloves or hands between inspections each colony and change clothes between apiaries.
- Clothes should be put in a sealed plastic bag in case any pests are on the clothing.
- Never poke your hive tool into any suspect disease brood cell.
- If a brood disease is found in a colony, change the hive tool or get a flame burning in your smoker and place your hive tool into the flame to heat up your hive tool. The heat from the smoker will kill any bacteria virus or fungal spores or reproductive bodies.
- Make sure that your smoker is in a safe place and be careful not to burn yourself.

Identifying diseases

Bees must be shaken off all brood combs, so you can see every cell on the frame as a disease or pest may only be in one cell and if present you should be able to find it. Shake the bees from the frame so the bees fall at the hive entrance or back into the box.

Specific diseases of the colony

You must be able to identify the signs but remember more than one brood disease may be present in the same colony. At different stages of the brood cycle, brood may be showing a different symptom of the disease, for example, American foulbrood has four signs and if one is observed then the disease is present.

Characteristics of healthy colony

It is vital you know what healthy brood looks like. Brood has three stages:

- *Egg*: The queen lays one white egg in the middle of the base of the cell. She lays the egg upright, and then over the next 72 hours the egg leans over to lie on its side on the base of the cell. The egg stage lasts 3 days.
- *Larva*: A healthy larva is pearly white in colour, glistens with a moist appearance and lies neatly coiled in the middle of the base of the cell. The larvae grow rapidly and can be observed with a white fluid – royal jelly – surrounding the younger larvae. The larvae stage for a worker is 8 days, for a queen 7 days and for a drone 11 days.
- *Pupa or sealed brood*: Brood cells from a healthy hive are sealed with a convex (raised) cap for both worker and drone brood. The capping is made from body hairs of worker bees and from wax taken from the cell wall. This makes the cappings a light brown colour and very different from the white cappings over honey. All cappings in the same area should be complete, with no holes and of a similar colour, except when the larva is in the process of being capped. The pupa stage for a worker is 10 days, a drone 10 days and a queen 6 days.

It is important to be able to identify healthy brood stages. Characteristics of healthy brood are:
- Healthy worker, queen, and drone larvae are pearly white in color with a glistening appearance.
- They are curled in a 'C' shape on the bottom of the cell and continue to grow during the larval period, eventually filling their cell.
- A healthy worker brood pattern is easy to recognize: brood cappings are medium brown in color, convex, and without punctures.
- Healthy capped worker brood normally appears as a solid pattern of cells with only a few uncapped cells; these may contain eggs, uncapped larvae, nectar, or pollen.

5.2 Diagnosis and Treatment

The diagnosis and treatment/prevention procedures to be followed for major diseases and parasites are described below.

5.2.1 American Foulbrood Disease

Diagnosis of AFB diseases

The two simple diagnosis techniques are the stretch test method and HOLTS milk test method.

1) Stretch test method

Diagnosis of AFB using stretch test method (Ritter and Akratanakul, 2006):
- A simple way of determining whether AFB caused the death of the brood is the stretch test.
- A small stick, match or toothpick is inserted into the body of the decayed larva and then gently and slowly, withdrawn.
- If the disease is present, the dead larva will adhere to the tip of the stick, stretching for up to 2.5cm before breaking and snapping back in a somewhat elastic way.
- This symptom called 'ropiness', typifies American foulbrood disease, but it can be observed in decaying brood only.

2) HOLTS milk test procedure

Another field test to confirm the presence of AFB can be conducted using powdered milk (MAAREC, 2011) (Fig. 10.2).

a) Preparing powdered milk
- Combine one teaspoon of the powdered milk with 100 milliliters (slightly less than half cup) of water and mix thoroughly.

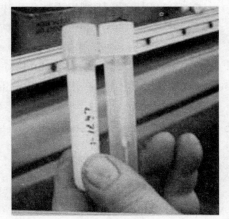

Fig. 10.2 HOLTS milk test
Source: MAAREC, 2011.

➢ Pour the milk into two small, clear, glass vials or other similar containers.

b) Collecting an AFB sample
➢ Collect a sample from the suspect AFB colony by opening a diseased cell and stirring the contents with a toothpick.
➢ Collect as much of the larval remains as possible on the toothpick and place in a clean container or wrap in foil.

c) Positive AFB sample
➢ Insert the previously collected AFB sample into one of the prepared vials.
➢ Do nothing with the second sample.
➢ Place both vials in a warm location for one hour.
➢ After one hour examine the samples.
➢ If the sample is positive, the vial with the AFB sample will become clear (Fig. 10.2 the right test tube).

d) Use the second sample for comparison

Treatment of AFB disease

Chemotherapeutic methods of controlling AFB involve the administration of antibiotics called *Oxytetracycline* (Terramycin) fed mixed with powdered sugar or sugar syrup. Drug treatment suppresses development of the AFB pathogen. However, treatment does not destroy spores. Terramycin is also effective in the treatment of European foulbrood. The treatment procedure is as follows (MAAREC, 2005; M. Nasr, 2014):

➢ Terramycin is a soluble powder is available in a 10g packet.
➢ The recommended dose is 200mg (one level teaspoon) of Terramycin mixed with 30g sugar powder, administered three times, at 4 to 5 - day intervals per colony.
➢ Spread the terramycin mixture over the end of the top bars in the hive body with the most brood located.
➢ Preventative treatments can be applied to colonies after honey supers are removed/harvested and again in the early flow period before putting honey supers on.

Caution: No drug should be fed when there is danger of contaminating the honey crop. Follow the directions carefully and do not overdose. Remember that spores may and often do survive feeding of these drugs and colonies may break down with disease when treatment is terminated.

5.2.2 European Foulbrood Disease

Diagnosis of EFB diseases

The most diagnostic characteristics of EFB are blotchy and twisted brood. Other symptoms include:

➢ The most significant symptom of EFB is the non-uniform color change of the larvae. They change from a normal pearly white to yellowish, then to brown, and finally to

grayish black; they can also be blotchy or mottled. Infected larvae lose their plump appearance and look undernourished.
- Their breathing tubes, or tracheae, are visible as distinct white lines.
- Larval remains often appear twisted or melted to the bottom side of the cell.
- Unlike larvae killed by AFB, recently killed larvae rarely pull out in a ropy string when tested using a stretch test.
- The dead larvae form a thin, brown or blackish brown scale that can be easily removed.
- EFB usually does not kill colonies, but a heavy infection can seriously affect population growth.

Treatment and preventive measures

The choice of an EFB control method depends on the strength of the infection, i.e. how many brood cells and combs are infested (Ritter and Akratanakul, 2006).

1) If the infection is weak

It is often sufficient to stimulate the hygiene behavior of the bees. Either they are placed at a good foraging site or they are fed with honey or sugar syrup. An even better result is achieved if the individual combs are sprayed with a thinned honey solution.

2) If the infestation is stronger

It makes sense to reduce the number of pathogens in the colony. Chemotherapeutic treatment measures such as antibiotics using the drug oxytetracycline should be used in serious cases. Follow directions for treatment using terramycin under AFB. Other measures include:
- Replacing the most infested brood combs with empty combs or healthy brood combs
- Re-queening can strengthen the colony by giving it a better egg-laying queen, thus increasing its resistance to the disease and interrupting the on-going brood cycle giving the house bees enough time to remove infected larvae from the hive.

5.2.3 Viral Diseases

Diagnosis

Ideally, the diagnosis of viral diseases is made using serological techniques. Since this is beyond the capability of most laboratories, diagnosis is usually made by observing symptoms in individual bees and, when possible, colony behavior (Shimanuki and Knox, 2000).

Viral diseases preventive measures

Viral diseases of honeybee are sac-brood disease, acute and chronic bee virus paralysis, deformed wing disease and black queen cell virus disease. Yet, there are no known direct chemotherapeutic (drug) methods of controlling any honey bee viruses. The message for

any viral disease, irrespective of the causative species, is the same (Lea, 2015). As a beekeeper, you can minimise the incidence and impact of viral diseases in your colonies by doing the following:
- Keeping strong, healthy colonies
- Monitoring regularly and treating for Varroa infestations
- Treating colonies for disease when Varroa populations are low (i.e. before the mites reach damaging levels)
- Adhering to good apiary hygiene practices
- Cleaning your equipment between each hive examination
- Good beekeeping management

5.2.4 Chalk-brood Disease

Diagnosis

- Diagnosis of chalk-brood disease involves is identification of chalk-brood mummies. The chalk-brood mummies are hard and resemble pieces of chalk when white.
- Dead larvae are chalky white and usually covered with fungus filaments (mycelia) that have a fluffy, cotton like appearance.
- These mummified larvae may be mottled with brown or black spots, especially on the undersides, because of the presence of maturing fungal fruiting bodies.
- Larvae that have been dead for a long time may become completely black as these fruiting bodies fully mature.
- Dead larvae (mummies) are often found in front of the hive, on the landing board, or in a pollen trap. In strong colonies, most of these mummies will be discarded by worker bees outside the hive, thus reducing the possibility of reinfection from those that have died from chalk-brood.

Treatment

The effective treatment method against chalk-brood disease is the use of Apiguard, which is mainly used for the treatment of Varroa mites. See the treatment procedure under the treatment of Varroa mite using Apiguard. Other important preventive measure is improving ventilation can help prevent chalk-brood.

5.2.5 Nosema

Diagnosis of Nosema infection

Seriously affected worker bees are unable to fly and may crawl about at the hive entrance or stand trembling on top of the frames. Damage to the digestive tract may produce symptoms of dysentery. Infected workers defecate in or on the outside of the hive or on frames. However, the clinical symptoms associated with Nosema infection can be seen with other types of colony conditions, so cannot be used to provide reliable diagnosis. Thus,

positive identification of this disease is through microscopic examination. A method to detect *Nosema* uses a light microscope to confirm presence of spores, as follows (Lea, 2015):
- Take a sample of 30 adult bees that are infected.
- Use a pestle and mortar, crush the sample of bees in a little water.
- Place a small drop of the resulting suspension onto a microscope slide and cover with a glass cover slip.
- Examine the suspension under the light microscope, magnification $\times 400$.
- The spores of both *N. apis* appear as translucent, greenish, rice shaped bodies. Determine the number of spore in a sample.

Treatment

The effective chemotherapeutic treatment method for treating *Nosema* is to feed the colony with Fumagillin B. The active ingredient of Fumagillin B is an antibiotic. Procedure:
- The recommended dose is to feed Fumagillin B (25mg active ingredient per litre of sugar syrup).
- Time of dosing: Preferably at a time when the colony is likely to encounter stress conditions, such as during a long rainy season.
- Infection level: Feed bees a mix of Fumagillin B with sugar syrup if there is a high level of infection ($>$1 million spores/bee).
- To ensure individual colonies receive the accurate dose of Fumagillin B, mix as per label instructions and apply using direct to colony feeding techniques (bag feeding, pail feeding, etc.).
- Protect your Fumagillin B medicated sugar syrup from direct sunlight when feeding bees.
- Replace 2 to 3 old brood combs (typically darker) from the brood box to reduce the level of *Nosema* and accumulation of acaricides in the wax.

Note: It is of the utmost importance that no medication be administered to colonies when there is a chance of contaminating the honey crop.

Acetic acid treatment

Fumigation of hive colony using 80% acetic acid could be used to treat Nosema. Procedures to be followed (Lea, 2015):
- 80% acetic acid is prepared by adding one part by volume of water to four parts of glacial acetic acid.
- In the bee spaces between each box in a stack, place a wad of cotton wool or other absorbent material, pre-soaked in 100mL of acetic acid.
- Obvious gaps in the stack must be sealed to prevent the fumes escaping.
- Leave the stacks undisturbed for one week.

> - The fumes kill *Nosema* but do not taint honey or pollen stored in the combs.
> - However, the acid fumes will corrode any exposed metal surfaces. Metal end spacers need to be removed before fumigation, and should be scalded in hot water containing washing soda.
> - The combs must then be aired before use, making sure that any combs containing honey are aired under cover, as a precaution against robbing.

5.2.6 Amoeba

Diagnosis

Malpighamoeba mellificae causes amoeba disease. Since this protozoan is found in the Malpighian tubules of adult bees, diagnosis can be made only by removing and microscopically examining the tubules for the presence of amoeba cysts. The cysts measure $5 - 8 \mu m$ in diameter. Malpighian tubules are long, threadlike projections originating at the junction of the midgut and the hindgut (Shimanuki and Knox, 2000). Procedure:
- They can be teased away from the digestive tract with a pair of fine tweezers.
- Once this is done, place them in a drop of water on a microscope slide and position a cover glass over them, applying uniform pressure to obtain a flat surface for microscopic examination.
- *M. mellificae* can be discerned using a high, dry objective and then changing to the oil immersion objective for more detail.

Treatment and control of amoeba/dysentery

Irrespective of its cause, there is no effective treatment for amoeba (dysentery), and cleansing flights that may alleviate the problem are entirely dependent on the weather (more common on warm, dry, windless days). However, there are steps that can be taken to reduce the risk of occurrence:
- As always, good husbandry and apiary management practices are vital in maintaining vigorous, healthy stocks that are more able to withstand infection.
- Dearth period feeding should be completed by the first week of honey flow period - allowing time for the colony to consume the sugar syrup.
- Avoid feeding bees with fermented honey or sugars of uncertain origin.
- Use only refined sucrose, table sugar or ready-made syrup mixtures (such as Ambrosia, Apisuc or equivalent).

5.2.7 Varroa Mites

Diagnosis methods

Monitoring and recognizing a Varroa infestation before it reaches a critical level is important. Diagnosis methods include examining drone cells, sampling bees using sugar shake, sticky board method and alcohol wash methods (Coffey, 2007; MAAFEC, 2011; Lea, 2015).

1) Examining drone brood

One technique to quickly assess the presence of Varroa mites is by examining drone brood.

- Uncap cells and remove and examine pupae, especially white drone pupae.
- Individual pupae can be removed using forceps, or many drone pupae can be removed at once using an uncapping fork.
- For examination: a small 10x hand lens will be helpful.
- If <5% of the brood is infested, the colony can be considered lightly infested, while >25% indicates a severe infestation and possible colony collapse (Coffey, 2007).

2) Sampling using sugar shake

The sugar roll technique is a quick, relatively easy sampling method to check for the presence and number of mites on the worker adults of a colony (Fig. 10.3).

Fig. 10.3 Sugar shake method
Source: MAAREC, 2011.

a) To collect an accurate sample (number of bees)

- Remove a frame covered with bees from the brood nest, taking care not to include the queen.
- Shake the bees into a plastic tub or cardboard box.
- Shake the tub to consolidate the bees into the corner.
- Scoop a half cup of bees using a half-cup measuring cup (a full half-cup measuring cup contains approximately 320 bees).
- Place the bees into a wide-mouth quart mason jar modified with a mesh hardware cloth top.

b) Counting the mites in a sample

- Add two to three tablespoons of powdered sugar to the bees in the jar. Vigorously shake the jar for about 30 seconds to distribute the sugar over the bees.

- Allow the jar to sit for approximately one minute.
- Then shake the loose sugar with dislodged mites out of the mason jar through the modified mesh cover onto a flat surface such as a cookie sheet, pie plate, or hive lid.
- Add more powdered sugar and re-shake until no additional mites appear after shaking.
- Count the number of mites.

3) Sampling using sticky boards

Another way of quantifying mite levels is by using a sticky board placed at the bottom of the hive.

- Coat a thick piece of paper (38cm × 30cm) cardboard (cartoon), using Vaseline as your sticky material to make sticky board.
- The sticky material must be covered with wire mesh screen elevated about 6mm off the sticky surface.
- Place the coated paper under a screen, on the bottom board for three days.
- Count the Varroa mites on the sticky board and divide by 3 to obtain an average of mites fall per day.
- Place boards in colonies for a minimum of three days to accurately calculate daily mite drop numbers.

4) Alcohol Wash

- Place 1/2 cup of bees (approximately 300 bees), from the brood chamber, into a container with alcohol so that the bees are completely immersed and shake vigorously. Pour bees onto a screen, over a white tub and vigorously rinse Varroa from bees with the remaining alcohol.
- Count the total number of Varroa in tub.
- Handheld, easy to use, commercially made mite shaker devices that give effective and fast results are available.
- Follow the directions provided with the shaking apparatus. Contact your local bee supply outlet for availability.

Treatment

Treat when Varroa levels are equal or greater than the following Table 10.1:

Table 10.1 Varroa levels in several monitoring methods for treatment

Monitoring methods	Number of Varroa mites
Drone brood	>20% of brood infected
Sugar shake	2 mites per 100 bees
Sticky board	10 mites drop per day
Alcohol wash	2 mites per 100 bees

Varroa mite (*Varroa destructor*) can be controlled/treated by many different treatments which may be subdivided into four main categories: chemical, biotechnical, organic acids and biological (Coffey, 2007).
- Chemical: e.g. Bayvarol
- Biotechnical: e.g. Mush floor type and drone brood trapping
- Organic acid: e.g. Formic acid and oxalic acid
- Biological: e.g. Apiguard and exomite

In general, the efficacy of chemical products is greater than non-chemical based products.

1) Bayvarol

This is a dearth period chemical treatment which should not be applied during foraging or before honey harvesting.
- It is a contact pesticide, killing mites only in the phoretic stage.
- The treatment period is six weeks, thus incorporating two brood cycles.
- Bayvarol strips are plastic strips impregnated with flumethrin.
- When treating a colony in a single brood box, four strips should be suspended in the spaces between the combs in the centre of the brood area. Two strips are sufficient when treating nuclei.

Bees crawl over the strips distributing the active ingredient throughout the hive. This product has $>95\%$ efficiency.

2) Apiguard

Apiguard is a natural product. It has a slow release gel matrix, ensuring the correct dosage of the active ingredient thymol. It has a proven high efficacy against the Varroa mite and is also active against both tracheal mite and chalk-brood disease. It is distributed around the hive by inhalation and contact. The percentage efficiency is estimated at 93%, but at ambient temperatures $<15℃$, efficiency is significantly reduced. Prior to administration, additional bee space should be created above the brood frames using an *eke*.

Procedure:
- Apiguard is administered by placing a 50g container over the brood area and two weeks later adding a second.
- Treatment should continue until the containers are empty or supers are about to be placed for the coming honey flow.
- An upturned feeder may also be used to provide bee space and hive roofs should be insulated to conserve heat.
- It is not recommended to apply Apiguard and feed simultaneously and mesh floors should be replaced with solid floors or at least closed off using an insert.

3) Drone brood trapping

In a queen right colony, the bees will naturally build drone comb and mites will be attracted to the developing larvae. Once the cells are sealed, the mites are trapped inside, and the comb can be cut off and removed from the hive. The shallow frame is placed back into the brood box so that the cycle may be repeated. Drone removal does not adversely affect colony health or honey production, thus may serve as a valuable component in an integrated pest management programme.

Large number of mites can be removed from an infected colony, without affecting the worker population using drone brood trapping. Mites are attracted to the drone brood for reproduction and thus a disproportionately large number of mites will be associated with the drone brood. During May-July, a shallow frame may be placed in the brood box, close to the brood area.

4) Mesh floors

In infested colonies, 39 – 50% of the mites which fall naturally from bees are alive and mobile and capable of re-infesting the colony. This live proportion of mite fall increases in warmer weather. Mesh floors, often referred to as Varroa floors comprise of #8 hardware screen (3 mesh/cm). The device can either be a standard bottom board whose solid floor has been replaced, or a rim (at least 10 – 20mm high) with screen made to fit between the brood box and standard floor.

This screen floor placed underneath the brood box prevents mites, knocked off by grooming, from returning to the hive and eliminates any contact between returning foragers and the expelled mites, thus preventing re-infestation. Additionally, studies have shown that mesh floors lower the percentage of the mite population residing in the brood cells and significantly increases the amount of sealed brood in a colony. Colonies may be placed on open mesh floors throughout the year.

5) Formic acid treatment

Formic acid is found as a natural component in honey and is effective against Varroa and tracheal mite. Formic acid fumigation as an alternative control has many advantages in that it is inexpensive, there is no documented resistance and it does not leave residues above natural levels in the honey. Formic acid treatment procedure (Kozak, 2013):

- Prepare 65% liquid formic acid.
- When using formic acid, make sure to seal all holes in the hive boxes except the main entrance, which must be left wide open (removing entrance reducers).
- Due to the high corrosive nature of this acid, irrespective of application method, it is essential that the user adheres to the recommended health and safety instructions.
- Wear protective gear (eye protection, chemical-safe gloves, long-sleeved shirt, and closed toe shoes).
- Have a container of water handy to wash off any splashes.
- Apply one 30 to 40 mL pad for double box or 15 to 20 mL pad for single box per hive. Place the pad on the top bars close to the brood area.
- During this period the vapor will be mixed in the hive air and travel down to the

bottom of the hive because the formic acid vapor is heavier than the air. This turns the hive into a fumigation chamber. As Varroa mites are exposed to the formic acid, they die.
- Leave treatment in for 21 to 30 days.
- The treatment is to be repeated up to six times and recommended at 3 to 5 - day intervals as per label instructions.
- The hive entrance must be open and unobstructed for all applications.

6) Oxalic acid treatment

Oxalic acid like formic acid is a naturally occurring acid. Residues do not accumulate in wax, and in honey are limited and toxicologically insignificant, assuming the beekeeper uses the acid according to the recommended health and safety instructions. To minimize the risk to the apiarist, protective clothing must be worn. Apply when monitoring indicates treatment is necessary.

Caution:

- Oxalic acid will not control Varroa mites in capped brood. Thus, use only when little or no brood is present; even it may damage bee brood.
- *Do not use* when honey supers are in place to prevent contamination of marketable honey.
- To prevent danger to yourself, wear protective gear (eye protection, chemical-safe gloves, long-sleeved shirt, and closed toe shoes). Have a container of water handy to wash off any splashes.

Procedure of trickling method:

- To completely dissolve oxalic acid dihydrate, use warm syrup (not hot) and agitate thoroughly. Dissolve 35g of oxalic acid dihydrate in 1 litre of premixed syrup made from a 1 : 1 sugar : water (weight : volume) mixture.
- Smoke bees down from the top bars.
- With an applicator (e.g. syringe), trickle 5mL of this solution directly onto the bees between the frames, apply on a cool day when the bees are clustered in the hive.
- The maximum dose is 50mL per colony whether bees are in nucs or single/multiple brood chambers.
- Under certain unfavourable conditions, e.g., weak colonies or unfavourable rainy conditions, this application method may cause some bee mortality or stress to the colony.

7) Exomite

This may be used at any time of the year. The product is applied at the hive entrance and delivers a very low application of Entostat powder and thymol.

As bees arrive at the hive entrance, they pass over the powder, collecting it on their bodies and carrying it into the hive. As they encounter other individuals, the preparation is disseminated throughout the hive by the bees themselves. Two consecutive treatments, each lasting 12 days are required for the control of the mite. The percentage efficacy is estimated at approximately 80%.

5.2.8 Tracheal Mites

Diagnosis of tracheal mites

Positive identification of tracheal mites is best done by dissection and microscopic examination of worker bee thoracic tracheae (Coffey, 2007; Lea, 2015).

1) Equipment and tools

The following tools are required for dissection and microscopic examination:
- A binocular dissecting microscope with a magnification up to $\times 40$, and a cool, concentrated light source
- A double dissecting needle
- A pair of fine-pointed steel forceps
- Small sheet of cork

2) Dissection procedure
- To determine the level of infestation in a colony, it is best to dissect and examine 30 freshly dead bees.
- Each recently-killed adult bee is laid on its back and pinned firmly to the cork by pushing the double needle at an angle through the thorax between the coxal joints of the second and third pairs of legs.
- The head and forelegs of the bees are firmly gripped between the forceps' blades and together detached cleanly from the body by a sharp pull of the forceps away from the bee and slightly upwards. The aperture now left in the thorax is focused under the microscope.
- The chitinous collar surrounding the aperture must now be removed to expose the thoracic breathing tubes. The collar is gripped with the points of the forceps where it is narrowest, and peeled off with a rotary motion of the closed forceps. The clean removal of the collar requires practice.

3) Characteristics of healthy and infested tracheae (under microscope):
- *Healthy*: The tracheae of uninfected bees are clear and colorless or pale amber in color (Fig. 36).
- *Slight infestation*: In a slight infestation, one or both tracheal tubes contain a few adult mites and eggs, which may be detected near the spiracular openings. At this stage, the tracheae may appear clear, cloudy, or slightly discoloured.
- *Sever infestation*: The tracheae of severely infested bees have brown blotches with brown scabs or crustlike lesions, or may appear completely black, and are obstructed by numerous mites in different stages of development. Feeding by the

mites damages the walls of the tracheae. Flight muscles in the bee's thorax also may become atrophied because of a severe infestation.

Treatment

For the treatment of tracheal mite, effective procedures are Apiguard and formic acid treatment. Follow the same guidelines described for the control of Varroa mites.

>>> SELF-CHECK QUESTIONS

Part 1. Multiple choices.
1. Which of the following is a viral disease of honeybee brood?
   A. European foulbrood diseases
   B. American foulbrood diseases
   C. Sac-brood disease
   D. All the above
2. Which of the following is a fungal disease of honeybee brood?
   A. European foulbrood diseases
   B. American foulbrood diseases
   C. Sac-brood disease
   D. Chalk-brood disease
3. Which of the following is a bacterial disease of honeybee brood?
   A. European foulbrood diseases
   B. Chalk-brood disease
   C. Sac-brood disease
   D. All the above
4. Which of the following is a protozoan disease of honeybee?
   A. Nosema
   B. Varroa mites
   C. Sac-brood disease
   D. All the above
5. Brood disease characterizes by affecting older sealed larvae or young pupae, dead brood become dull white then brown in color with soft, sticky and ropy consistency that uniformly lies flat and adheres on lower side of cell.
   A. European foulbrood diseases
   B. American foulbrood diseases
   C. Sac-brood disease
   D. All the above
6. Brood disease characterizes by affecting young unsealed larvae, dead brood become dull white then yellowish to dark in color with watery to pasty consistency that lies twisted in a cell.
   A. European foulbrood diseases
   B. American foulbrood diseases
   C. Sac-brood disease
   D. All the above
7. Brood disease characterizes by affecting sealed brood/pupa, dead brood become grayish then brown to dark in color with watery and granular tough skin or sac consistency with brittle texture or form that lies head curled up in a cell.
   A. European foulbrood diseases
   B. American foulbrood diseases
   C. Sac-brood disease
   D. All the above
8. Fungal disease of adult bees characterizes by affecting intestinal tract, worker bees

are unable to fly and may crawl about at the hive entrance or stand trembling on top of the frames, excreta on combs or entrance boards, and a pile of dead bees on the ground in front of the hive.

  A. Nosema   B. Varroatosis
  C. Tracheal mites   D. Paralysis

9. Brood disease characterizes by affecting older larvae, the dead larvae swell to the size of the cell and are covered with the whitish mycelia of the fungus. Subsequently, the dead larvae mummify, harden, shrink and appear chalklike.

  A. European foulbrood diseases   B. Chalk-brood disease
  C. Sac-brood disease   D. American foulbrood disease

10. Which of the following disease – causing agents does not have a direct Chemotherapeutic (drug) treatment method?

  A. Bacteria   B. Protozoa   C. Fungus   D. Virus

11. The most appropriate diagnosis procedure to identify AFB disease is?

  A. HOLTS milk test   B. Sugar shake method
  C. Sticky board method   D. Alcohol wash method

12. Which of the following is not used to monitor and diagnose Varroa mite?

  A. HOLTS milk test   B. Sugar shake method
  C. Sticky board method   D. Alcohol wash method

13. The most effective treatment method (drug) to control AFB & EFB diseases is?

  A. Apiguard   B. Bayvarol
  C. Terramycin   D. Fumagillin B

14. The most effective treatment method (drug) to control Nosema diseases is?

  A. Apiguard   B. Bayvarol
  C. Terramycin   D. Fumagillin B

15. Which of the following treatment method effective against Varroa mite, tracheal mite and chalk brood disease?

  A. Apiguard   B. Bayvarol   C. Formic acid   D. Oxalic acid

Part 2. True or False.

1. In HOLT's milk test, if the sample is positive, the vial with the AFB sample will become clear.

2. In HOLT's milk test, if the sample is positive, the vial with the AFB sample will have milk color.

3. Breathing tubes of a healthy bee have a uniform creamy-white appearance.

4. The tracheae of severely infested bees with tracheal mite have brown blotches with brown scabs or crustlike lesions, or may appear completely black.

5. Avoid feeding bees with fermented honey or sugars of uncertain origin to prevent amoeba.

Part 3. Match Column 'A' with the appropriate words/phrases from column 'B'.

MODULE 10
CARE FOR THE HEALTH OF HONEYBEE COLONY

A	B
1. *Paenibacillus larvae*	A. Causative of European foulbrood diseases
2. *Melissococcus pluton*	B. Causative of American foulbrood diseases
3. *Morator aetotulas*	C. Causative of bee amoeba
4. *Malpighamoeba mellificae*	D. Bee lice
5. Varroa treatment	E. Causative of sac-brood disease
6. *Galleria mellonella*	F. Small hive beetle
7. Ants control measure	G. Ants
8. Poisonous plants	H. Wax moth
9. *Aethina tumida*	I. Formic acid
10. *Dorylus fulvus*	J. Solanaceae, Acanthaceae &. Euphorbiaceae
11. Viral disease	K. Birds, honey badger and lizards
12. *Aspergillus flavus*	L. Varroa control measure
13. *Braula coeca*	M. Bee paralysis
14. Drone brood trapping	N. Putting hive stand in tin with car oil
15. Predators	O. Causative agent of stone brood disease

Part 4. Discussion.

1. Discuss the causes, symptoms and control measures of robbing in bees.

2. Discuss the signs/symptoms and control measures of ant attacks.

3. Discuss the signs/symptoms and control measures small hive beetles.

4. Discuss the signs/symptoms and control measures of wax moth.

5. Describe and discuss general guidelines of honey bee diseases, pests and control measures.

>>> REFERENCES

Begna D, 2007. Assessment on the effect of ant (*Dorylus fulvus*) on honeybee colony (*A. mellifera*) and their products in West and South West Shewa Zone: Ethiopia [J]. Ethiopian journal of animal production, 7 (1): 12-26.

Begna D, 2014. Occurrences and Distributions of Honeybee (*Apis mellifera Jemenetica*) Varroa Mite (*Varroa destructor*) in Tigray Region, Ethiopia [J]. J Fisheries Livest Prod, 2: 126.

Begna D, 2015. Honeybee diseases and pests research progress in Ethiopia: A review [J]. African Journal of Insect, 3 (1): 93-96.

Cramp D, 2008. A practical manual of bee keeping [M]. Spring Hill House, United Kingdom.

Gidey A, Mulugeta S, Fromsa A, 2012. Prevalence of Bee Lice *Braula coeca* (Diptera: Braulidae) and other perceived constraints to honey bee production in Wukro Woreda, Tigray Region, Ethiopia [J]. Global Veterinaria, 8 (6): 631-635.

Segeren P, 2004. Beekeeping in the tropics [M]. 5th ed. Digigrafi, Wageningen, the Netherlands.

Tesfay H, 2014. Honey Bee Diseases, Pest and Their Economic Importance in Ethiopia [J]. International Journal of Innovation and Scientific Research, 10: 527-535.

Yohannes A, Bezabeh A, Yaekob B, et al., 2010. Ecological distribution of honeybee chalkbrood disease (*Ascosphaera apis*) in Ethiopia [J]. Ethiopian J Ani Prod, 9 (1): 177-191.

MODULE 11:

RECORD KEEPING

>>> INTRODUCTION

Keeping accurate records is important. Records help you remember what you did and evaluate the success of your work. They also help you keep track of how much time and money you are spending on your beekeeping project (Greg and Natalie, 2015). Both production and financial records are important to the efficient management of today's farm business. When such information is accurately maintained and categorized, it can be used to produce useful decision-making information. Generally, there are different record formats, which are must applicable in apiary to increase beekeeping production and productivity. Therefore, this module is developed to cover the necessary information to recognize types of records and how to develop record format for beekeeping.

1 RECORD - KEEPING PRACTICES IN BEEKEEPING

Record your observations and actions because of observations per hive. This would necessarily mean that your hives should be numbered/labeled to avoid confusion. In addition, a record about the date of harvest, produce acquired per hive in kilogram and the monetary value of the produce should be kept. To avoid confusion, make colony records and operational records. Operational records are mostly concerned with expenditure and cash flow. With this type of record, you would be able to know whether you are making profit or running a loss (Gokwe vocational training centre, 2011). A serious beekeeper studies his bees and records their behavior seasonally. The behavior of bees different with different seasons; there is a season that encourages swarming. Bees produce more honey during a certain season. Certain types of bee fodder plants bloom during specific seasons. Bees require supplementary food during a certain season.

Beekeepers should customize their record-keeping system to their own operations, using different record format and considering principle of record keeping (Canadian food inspection agency, 2013). Principles of record-keeping are:

➢ Complete records in 'real time' whenever possible. Recording after-the-fact (from

memory) can often lead to errors.
- Have records that are as accurate as possible. Unconfirmed diagnosis or suspicion of a pest should be identified as such.
- Be aware that errors in entering information should be struck-through, dated, and initialed, rather than erased or otherwise obscured.
- Know that dated and properly identified digital camera images are a useful supplement to written records.
- File all documents such as receipts, invoices, diagnostic reports, and permits in a secure location.

2 TYPES OF RECORD IN BEEKEEPING

Different types of records are maintained in beekeeping activities. The types of apiary records and their developed formats are discussed in the subsequent sections.

2.1 Breeding Record

One of the main problems with keeping records is deciding where the record stays. Is it with the colony or the queen? The simple answer might be if the purpose is for colony records, they should stay with the colony, if it is "breeding records" then with the queen. In breeding records, the following information should considered:
- Date
- Is the queen laying?
- Is the queen clipped and marked?
- Are there any queen cells?

If the queen is not laying when she should be, we need to find out why and deal with it if needed. If she is not clipped and marked when she was the last time we saw her, then something has happened. If there are queen cells, we need to deal with them and think what is done next. Characteristics used for selecting queens to breed from or to cull are temper and calmness on the comb (Table 11.1).

Table 11.1 Breeding record format

Hive	Queen stock								
	Queen identifier (mark)	Supplier name	Queen installation year/month	Source or strain (parent colony)	Queen accepted and replacing rejected queen	Queen mother ID	Drone colony ID	Advantage (e.g. hygienic behavior)	Disadvantage (e.g. swarming, aggressive behavior)

2.2 Colony Health Record

The first line of defense is protecting your colonies from diseases, parasites, and pests. You must detect and recognize early symptoms of these problems. Failure to recognize problems early can lead to decreased productivity and weak and even dead colonies. Honeybees are classified as a food-producing animal and therefore beekeepers must comply with the veterinary medicines regulations. Treatment given and presence of diseases, parasites, insect pests should be registered (Table 11.2 and Table 11.3).

Table 11.2 Treatment given when acquire record format

Hive ID	Treatment given when acquire					
	Type	Start date	End date	Who administered	Reason	Efficacy

Table 11.3 Record format for the problem due to presence of diseases, parasites, insect, and pests

Apiary site	ID number	Date of diagnosis	Symptom diseases, pests and parasite observed	Observed diseases, pests and parasite	Treatment		Remark
					Date	Types of treatments	

2.3 Colony Record

Colony record is important to ensure the number of colony we have and to identify productive and unproductive colonies (Table 11.4).

Table 11.4 Colony record format

Apiary site	Unit	Quantity	Days and month	Types of hive			Remark
				Traditional	Transitional	Modern	

2.4 Production Record

Production records are items that relate to quantities of inputs and levels of production by enterprise and/or by resource type. They consist of honey produced, wax, pollen, propolis, royal jelly colony multiplication, bee colony loss, etc. Keeping and analyzing accurate production records are important and essential aspects of farm management (Table 11.5).

Table 11.5 Production record

Apiary site	Hive ID	Year and date harvest	Type of product				Unit kg	Quantity	Remark
			Honey	Wax	Propolis	Pollen			
Total production									

2.5 Feed Record

For healthy bees, ensure access to a good quality carbohydrate source (nectar or supplement), pollen or pollen substitutes (protein, lipids, vitamins and minerals), supplementary pollen feed should be free from disease and access clean water. Monitor and provide supplemental feed and/or water, as required (Table 11.6).

Table 11.6 Feed records

Apiary site	Hive ID	Date	Types of feed	Unit kg	Quantity	Remark
Total feed						

2.6 Colony Source Record

In our country, bee colonies are collected from the swarming, splitting, purchased from market and queen rearing farmers (Table 11.7).

Table 11.7 Colony source record format

Hive	Type Colony source	History of colony			Remark
		Acquisition year/month	Source or strain (e.g. parent colony or name of supplier)	Behavior/health observation on receipt	
	P=package bees; N=nucleus colony; SW=swarm capture; SP=split				

2.7 Financial Record

Keeping and analyzing farm financial records are essential for the efficient management of a farm business. Accurate records and resulting analyses help farmers make financial and production decisions, comply with tax laws and other governmental regulations and support loan applications. Record keeping system can be practiced through traditional hand record-keeping or computerized record-keeping systems. Developing and using a farm financial record-keeping system will allow the farm manager to make more the correct decisions affecting the profitability of the farm income and expenses (Table 11.8).

Table 11.8 Farm income and expense record keeping format

Date	Description	Income		Expense				Profit
		Honey	Wax	Feed supplement	Labor	Materials	Other	
Total								

2.8 Materials Record

There are different types of materials and facilities that needed for beekeeping. Beekeeping equipment and facilities include beehives, honey extractor (centrifugal), honey presser, uncapping fork, honey storage, honey strainer, chisel (beekeeper's tool), knives, brushes, smoker, water sprayer, personal protective equipment, transportation facilities and other, so this all must be documented properly (Table 11.9).

Table 11.9 Materials record format

Apiary site	Materials type	Unit	Quantity	Date of purchased	Used or new

2.9 Human Resource Record

Employers, owner and other concerned bodies in the farm directly or indirectly participant in a given apiaries of beekeeping activities for the fruit full of outcomes of the farm. Due to this reason, in the farm the profile of employers, owner and other concerned bodies must be documented properly (Table 11.10).

Table 11.10 Human resource record format

Type	Number in sex		Title	Organization	Address	Telephone number	Email	Website	Date contact
	Male	Female							
Staff									
Extension specialist									
Suppliers									
Veterinarian									
Customer for custom product									
Driver									
Visitor									

>>> SELF-CHECK QUESTIONS

Part 1. Match the items recorded with types of record from column B to column A and write the letter you choice on the space provided for you.

A	B
1. Mother and sire identification number	A. Health record
2. Presence of diseases and treatment	B. Bees product record
3. Honey, wax and propolis	C. Financial record
4. Vehicles and house	D. Visitor record
5. Employer	E. Personal record
6. Hive and personal protective equipment	F. Facilities record
7. Supplement feed and nectar flow	G. Materials record
8. Income and expense	H. Colony disposition record
	I. Breeding record
	J. Feed record
	K. Baiting hive

Part 2. Write the correct answer accordingly.

1. Write advantage of record keeping.
2. List at least five types of record in beekeeping.
3. Write the principles of record-keeping in beekeeping.

>>> GLOSSARY

Absconding: Leaving the comb by the bee to unknown destination, due to unsuitable conditions

Apiarist: A person who is employed as a beekeeper or to aid beekeeping activities of others

Apiary: Place where beehives are kept; a secure place to carry out bee keeping activity

Apiculturist: A person with formal training in the study of bees, sometimes including bee species other than honey bees

Apitherapy: Using bee products to cure certain disease

Artificial swarming: Method of avoiding swarming by splitting strong colony into two

Bait hive: A hive left without bees in the hope that a swarm of bees will find it and move in, so that the beekeeper can obtain a new colony or catch a swarm leaving from one of the hives in the beekeeper's apiary. It is empty hive that is set up to attract swarm.

Bee brush: Used for brushing bees off the frame while inspection, a feather can also be used

Bee escapes: These are small device fitted to inner caver/swarm board. They provide only one-way movement.

Bee plants: The plants which provide food resource/forage source to bee

Bee races: Indicate geographical location of bees originating homeland and defines their traits

Bee space: A distance of 0.7 to 1.0cm which the bees naturally maintain between adjacent combs

Bee sting: Modified ovipositor, well developed in worker, present in queen, absent in drone. It is organ of defense.

Bee toxicity: The toxic effect caused by pesticide and poisonous plants to bees

Bee venom: A product of sting apparatus as defensive mechanism

Beehives: Enclosed structure in which some honeybee species live and rear their young

Bottom bar: The bottom piece of a frame, usually wooden

Bottom board: The bottom part of a hive, which supports the rest of the hive

Brace comb: The workers will connect the frame if they are not spaced properly.

Brood chamber: Portion of the comb where young ones are raised

Burr comb: Comb built not in the frames but between frames or between frames and the inner wall of a hive body, usually because the frames are too far apart

Cappings: The workers cap over drown-out cells with thin wax capping, to protect larva & honey.

Cell: Hexagonal structure built of wax by worker bee. It is the basic unit of the comb and used to store pollen, honey and rising of young ones.

Colony: The bees which live together in a cohesive society, including adult bees and brood but not necessarily including the hive they inhabit

Colony collapse disorder (CCD): A condition in which many adult bees suddenly disappear from the hive, for no obvious reason, but now apparently caused by a combination of mites and other factors

Comb: The wax structure made of many hexagonal cells in a hive, which the bees use for storing honey and pollen, and for rearing brood

Comb foundation sheets: Made of beeswax, commercial wax or plastic, using a comb foundation mill. Fixed to the farms, the hexagonal impressions help in speeding up comb building activities of the honeybee.

Dance: Type of physical communication, exhibited by worker bee to convey the availability of food resource in the field, to the fellow workers

Decapper: Used to remove wax seal of honey cells to make easier for extraction of honey by spinning

Dividing board: A solid brood shape frame, used to divide/limit brood frames

Drawing out comb: The process by which bees construct their comb by adding wax and forming hexagonal cells on the foundation

Drifting: The movement of bees from their own hive to another hive nearby, usually because they aren't orienting to their own hive very well

Drone: The male honey bee. They are recognizable by bearing much bigger than the workers.

Emergency queen cells: When the colony loses a queen, the workers build emergency queen cells.

Entrance block: Hives have entrance blocks designed to completely close the entrance, or to limit it to prevent robbing.

Extract: To spin honey frames rapidly so that the honey is thrown out of them, prior to the filtering and bottling of the honey

Extractor: A large cylindrical device used to extract honey from frames by spinning them rapidly, working much like a centrifuge

Fence: Prevents disturbance to beehive and for protection

Feral bees: Bees not kept by a beekeeper but living without management, for example, in a hollow tree

Forage: The food resource (pollen, nectar etc.)

Forager: Worker which gather forage

Foul brood: Disease affecting the brood, bad smell of the brood caused by fungi/virus

Frame: A hive part which holds the comb and can be removed from the hive for examination or honey extraction

Frame feeder: Hollow brood frame which can be filled with sugar syrup and left in the hive to feed nucleus or hived swarm

Gloves: Protective gear for hand, made of plastic, asbestos cloth, cotton cloth etc.

Gum hives: Hives kept by beekeepers in hollow logs or other non-standard equipment, also called 'bee gums'

Harvesting: Collecting pollen, honey, royal jelly or other bee products

Hive: The shelter used by bees, such as a wooden box or hollow tree; may include the bees

Hive beetle: The small hive beetle

Hive bodies: The wooden boxes used as part of the hive

Hive entrance: Open place usually at the bottom side of the hive for entry and exit of worker bees

Hive stand: Stand to protect enemies, rain water etc., to mount hive to suitable position

Hive tool: Tool which the beekeeper uses for freeing super and farms from which are propelled the hive

Hiving bees: Putting a swarm or package of bees into a hive

Honey: Regurgitated, fermented nectar. Used for feeding young ones, preparing wax, used as food and medicine by human beings

Honey bound: A hive condition in which so much honey has been stored by the bees. They do not have room to store more honey or rear brood.

Honey flow: Means the peak nectar flow, which the bees flat out collecting and storing

Honeydew: A liquid like honey but made by bees from the secretions of homopterous insects like aphids, instead of from flower nectar

Inner cover: The cover to a hive which rests directly over the hive bodies larva

Knife: Sharp edge instrument used to separate comb from the frame. May be electrically heated and used to decap honey cells

Langstroth: An American design hive, probably the most popular in the western world

Larva (larvae, plural): The worm-shaped immature form of the bee, which hatches from the egg

Laying worker: Worker bees that lay eggs, a problem in colonies that have been queenless for a long time

Manipulation cloth: Canvas cloth on an alloy frame used to cover an open brood box or super when examining frames

Melliferous flora: Bee plants visited by *A. mellifera*

Mouse excluder: Nails or zinc strip attached to hive entrance to avoid mouse entering hive

Nectar: The sweet liquid produced by flowers and collected by bees, and then made into honey by the bees

Nectar sac: Portion of digestive tract present in worker bee to carry nectar from flower to comb

Nuc (pronounced 'nuke', short for nucleus): A very small hive, usually for sustaining queen bees as they mature and mate or as hive that will eventually grow to full strength

Nucleus: A young colony of bees covering a few frames

Nucleus box: A small/half size brood box used to house a nucleus

Observation hive: A small hive with transparent sides, usually kept indoors for close viewing of bee activities

Outer cover: The cover to a hive that fits over the inner cover and the rest of the hive; the type which overlaps the hive body below is called a telescoping cover

Package bees: A colony of bees including the queen, in a screened box purchased from a commercial producer

Pepper pot brood: Brood rejected by the workers due to some fault

Pheromone: A chemical used by an animal to communicate with one of its own kind; very important to honey bees and other social insects

Pollen: Very small dust-like grains produced by flowers, these are the male germ cells of plants.

Pollen basket: Structure present in tarsal region of hind leg of worker bee to carry pollen grains

Pollen bound: A hive condition in which so much pollen has been stored in the brood nest by the bees, they do not have room for rearing brood

Pollen chamber: The portion of the comb where pollen is stored

Pollen trap: A device placed in front of the hive entrance to collect pollen grain from the foragers

Predators: Animals which feed on bees

Propolis: A sticky and resinous substance made by bees to fill in cracks and openings in their hive, and as an antibiotic to protect the bees from diseases

Pupa: The immature form of a bee as it transforms from a larva to an adult

Queen bee: Fertile female with life span of 2 - 3 years. Control the morphological and physiological development of workers. Only individual in the colony capable of laying eggs

Queen cage: Metal wire tube/cylinder with cork at both ends. Used to isolate queen from colony for management and for transporting/calling queens etc.

Queen cell: Constructed at the edge of the comb. Used to raise the queen

Queen excluder: Metal or plastic mesh restrict the movement of the queen. Usually placed between two compartments of hive

Queen gate: Perforated metal strip, to confine the queen inside the hive

Queen substance: Pheromone secreted by queen to maintain harmony in the colony

Re-queening: Technique designed to prevent bees swarming by introducing a new young queen

Ripening: Process of fermenting nectar into honey by worker bee

Robber bees: During forage shortage, the bees go to the nearest source for honey/other colonies

Royal jelly: Special type of feed prospective by worker to feed proactive queen larvae all her life and young larvae up to 3 days

Rustic hives: Hives kept by beekeepers in hollow logs or other non-standard equipment; also called 'gum hives' or 'bee gums'

Sac-brood disease: A disease of bee larvae that causes the larva to die and then resemble a sack of liquid

Scattered brood: An irregular brood pattern (see 'shot brood')

Shot brood: An irregular pattern of brood on a comb that shows many empty cells, like the pattern made by a shotgun on a target; caused by a queen with fertility problems or by laying workers stand —a local term for hive

Super: One of the upper hive bodies, which contains the honey frames

Supersedure: The process in which the original queen is gradually replaced by one of her daughters, usually because she is old or diseased

Swarm: A colony of bees not in a hive which has not yet found a place to live; also, the bees purchased in a package

Swarming: A process of division of colony/multiplication/resulting from multiple queens or overcrowding. Usually coincides with peak forage flow season

Top bar: The top piece of a frame, usually wooden

Tracheal mite: A highly destructive microscopic mite that parasitizes bees

Travelling box: Nucleus box designed for carrying bees from one place to other

Varroa mite: A highly destructive external mite that parasitizes bees

Wax: Secreted by worker to build comb

Wax moth: Pest, which feeds on wax, makes tunnels and the comb becomes weak. Finally results in absconding

Wild bees: Either honey bees not kept by a beekeeper (see 'feral bees'), or other species of bees such as bumblebees or orchard bees

>>> ANSWER KEY OF SELF-CHECK QUESTIONS

MODULE 1

Part 1. Multiple choice				Part 2. Matching			
1	D	6	A	1	J	6	F
2	C	7	A	2	B	7	G
3	A	8	D	3	A	8	H
4	D	9	B	4	D	9	I
5	C	10	C	5	E	10	C

MODULE 2

Part 1. Multiple choice			
1	A	6	D
2	D	7	C
3	D	8	A
4	B	9	A
5	B		

MODULE 3

Part 1. Multiple choice								Part 2. Matching						Part 3. True/False	
1	D	6	B	11	D	16	C	21	A	1	H	6	G	11	F	1	False
2	D	7	D	12	D	17	A	22	B	2	K	7	C	12	D	2	True
3	C	8	B	13	C	18	C	23	C	3	J	8	N	13	O	3	False
4	A	9	A	14	A	19	D	24	B	4	I	9	A	14	L	4	False
5	D	10	A	15	D	20	A	25	D	5	M	10	B	15	E	5	False

MODULE 4

Part 1. Multiple choice		Part 2. Matching part	
1	C	1	E
2	B	2	B
3	A	3	D
4	C	4	A
5	D	5	C
6	C		

MODULE 11
RECORD KEEPING

MODULE 5

Part 1. Multiple choice					Part 2. Matching				
1	D	6	D	1	E	6	I
2	A	7	D	2	D	7	F
3	D	8	B	3	C	8	J
4	D	9	B	4	H	9	G
5	C	10	D	5	A	10	B

MODULE 6

Part 1. Multiple choice					Part 2. Matching				
1	D	6	B	1	D	6	E
2	D	7	C	2	E		
3	E	8	E	3	F		
4	B	9	E	4	B		
5	B	10	A	5	A		

MODULE 7

Part 1. Multiple choice	
1	D
2	D
3	C
4	D
5	E

MODULE 8

Part 1. Multiple choice					Part 2. Matching				
1	B	6	A	1	I	6	B
2	B	7	F	2	C	7	D
3	B	8	D	3	J	8	F
4	B			4	E	9	H
5	D			5	G	10	A

MODULE 9

	Part 1. Multiple choice		
1	C	6	B
2	D	7	A
3	B	8	C
4	C	9	D
5	A	10	A

MODULE 10

Part 1. Multiple choice						Part 2. Matching						Part 3. True/False	
1	C	6	A	11	A	1	B	6	H	11	M	1	True
2	D	7	C	12	A	2	A	7	N	12	O	2	False
3	A	8	A	13	C	3	E	8	J	13	D	3	True
4	A	9	B	14	D	4	C	9	F	14	L	4	False
5	A	10	D	15	A	5	I	10	G	15	K	5	True

MODULE 11

	Matching part		
1	I	6	E
2	D	7	G
3	A	8	H
4	B	9	J
5	F	10	C

图书在版编目（CIP）数据

蜜蜂养殖＝BEE KEEPING：英文/埃塞俄比亚农业职业教育系列教材编委会组编．—北京：中国农业出版社，2018.4
 ISBN 978-7-109-23805-3

Ⅰ.①蜜… Ⅱ.①埃… Ⅲ.①蜜蜂饲养－饲养管理－英文 Ⅳ.①S894

中国版本图书馆 CIP 数据核字（2017）第 327504 号

中国农业出版社出版
（北京市朝阳区麦子店街 18 号楼）
（邮政编码 100125）
责任编辑 肖 邦 黄向阳

中国农业出版社印刷厂印刷 新华书店北京发行所发行
2018 年 4 月第 1 版 2018 年 4 月北京第 1 次印刷

开本：787mm×1092mm 1/16 印张：15.25 插页：6
字数：346 千字
定价：90.00 元
（凡本版图书出现印刷、装订错误，请向出版社发行部调换）

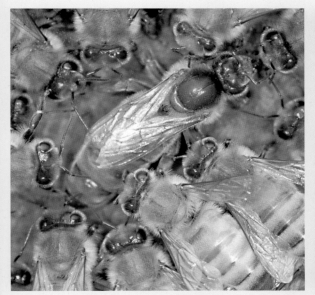

Fig. 1 *Apis mellifera*

Fig.2 *Apis cerana*

(Courtesy of Zhongyin Zhang)

Fig.3 Carniolan honeybee
(Courtesy of apiculture science institute
of Jilin Province, P. R. China)

Fig.4 Africanized queen
(Source: Glem Apiaries)

Fig.5 Three types of honey bees: queen, worker, and drone
(Source: www.dkimages.com)

Fig.6 Queen bee surrounded by worker bees
(Courtesy of Zhongyin Zhang)

A
(Courtesy of Zhongyin Zhang)

B

Fig.7 Worker bee collecting nectar and pollen grain
A. Collecting nectar   B. Collecting pollen

Fig.8 Pupa of bees
(Courtesy of Zhongyin Zhang)

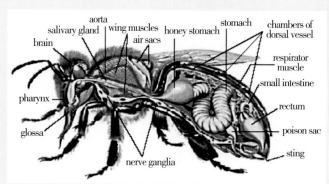

Fig.9 General internal organs of honeybees
(Source: Hennery, 2009)

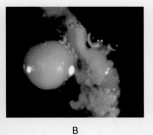

Fig.10 Reproductive organ of queen bees
A. Ovary of a laying queen   B. Spermatheca
(Courtesy of Zachary Huang)

Fig.11 Zander hives
(Source: Agarfa ATVET College Apiary)

Fig.12 Materials for foundation sheet making
A. Wax   B. Frame wire   C. Casting mould   D. Frame
E. Transformer/embedder
(Source: Agarfa ATVET College Apiary)

Fig.13 Set of hive tool (chisel and bee brush)
(Source: Agarfa ATVET College Apiary)

Fig.14 Honey processing equipment
A. Closed bucket   B. Uncapping fork and knife   C. Centrifugal extractor   D. Honey settling tank
E. Honey press   F. Refractometer   G. Honey jar

(Source: Agarfa ATVET College Apiary)

A
(Source: www.xtec.es)

B
(Courtesy of Shenglu Chen)

Fig. 15 Honeybee flora
A. Sunflower (*Helianthus annuus*)   B. Acacia species

A
(Courtesy of Zhongyin Zhang)

B
(Source: Agarfa ATVET College Apiary)

Fig.16 Fixing foundation sheets into frames
A. A Fixed wire with frame   B. Fixed foundation sheet with frame

Fig.17 Personal protective equipment
(Source: Agarfa ATVET College Apiary)

Fig.18 Newspaper method
(Courtesy of Zhongyin Zhang)

Fig.19 Pure wax
(Courtesy of Zhongyin Zhang)

Fig.20 Propolis
(Courtesy of Zhongyin Zhang)

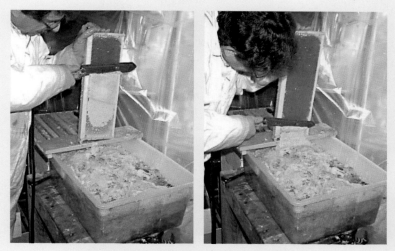

Fig.21 Cutting the comb using uncapping knife

Fig.22 Wax extraction procedure using hot water and squeezing method

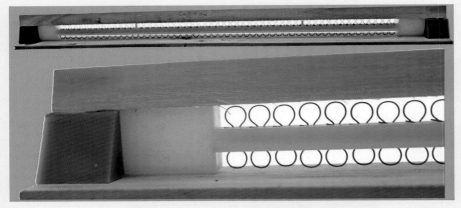

A

B

Fig.23 Pollen trap

(Courtesy of Zhongyin Zhang)

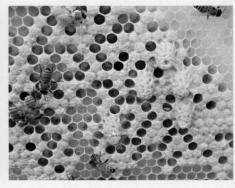

Fig.24 Queen cells
(Courtesy of Zachary Huang)

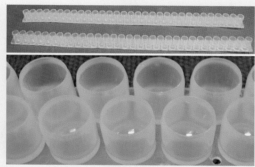

Fig.25 Plastic queen cells
(Courtesy of Zhongyin Zhang)

Fig.26 Grafting tools

(Courtesy of Zhongyin Zhang)

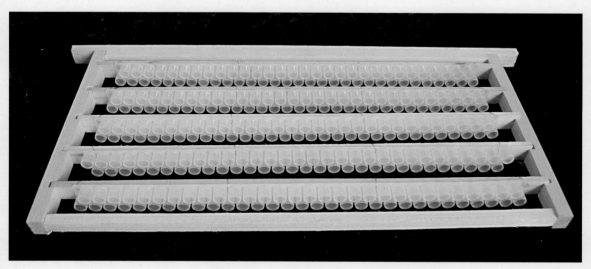

Fig.27 Artificial queen cells fixed to cell holder bars

(Courtesy of Zhongyin Zhang)

Fig.28 Grafting the larvae to artificial queen cells
(Courtesy of Zhongyin Zhang)

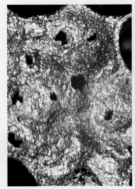

Fig.29 American foulbrood disease
(Courtesy of Zachary Huang)

Fig.30 European foulbrood disease
(Courtesy of Zachary Huang)

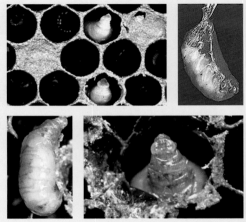

Fig.31 Sac-brood disease
(Courtesy of Zachary Huang)

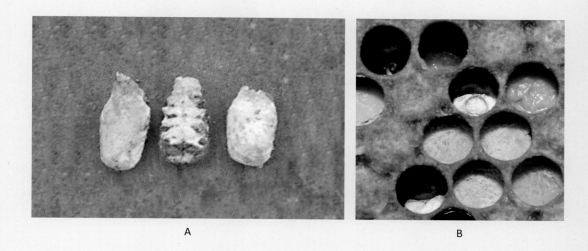

Fig.32 Chalk-brood disease

(Courtesy of Zhongyin Zhang)

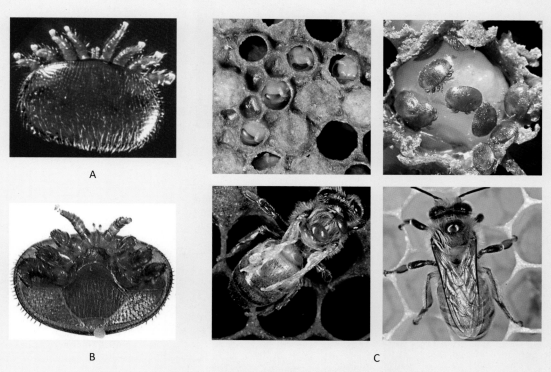

Fig.33 Varroa mite and symptoms

(Courtesy of Zhongyin Zhang and Zachary Huang)

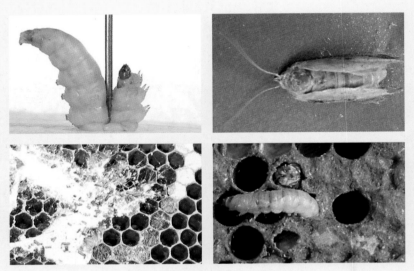

Fig.34 Wax moth
(Courtesy of Zhongyin Zhang)

A B

Fig.35 Pests and predators
A. Hornet B. Spider

(Courtesy of Zhongyin Zhang)

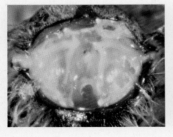 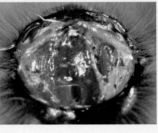

Fig.36 Characteristics of healthy and infested tracheae
(×40 under microscope)
A. Breathing tubes of a healthy bee – note a uniform, creamy-white appearance B. Breathing tubes of an infested bee – note patchy discolouration

(Source: Coffey, 2007)

A B